Z

DONALD G. McNEIL JR.

ZIKA

THE EMERGING EPIDEMIC

W. W. Norton & Company
Independent Publishers Since 1923
New York • London

Printed in the United States of America
First Edition

For information about permission to reproduce selections from this book, write to Permissions, W. W. Norton & Company, Inc., 500 Fifth Avenue, New York, NY 10110

For information about special discounts for bulk purchases, please contact W. W. Norton Special Sales at specialsales@wwnorton.com or 800-233-4830

Manufacturing by RR Donnelley Harrisonburg
Production manager: Julia Druskin

ISBN 978-0-393-60914-1
ISBN 978-0-393-35396-9 (pbk.)

W. W. Norton & Company, Inc.
500 Fifth Avenue, New York, N.Y. 10110
www.wwnorton.com

W. W. Norton & Company Ltd.
Castle House, 75/76 Wells Street, London W1T 3QT

1 2 3 4 5 6 7 8 9 0

Contents

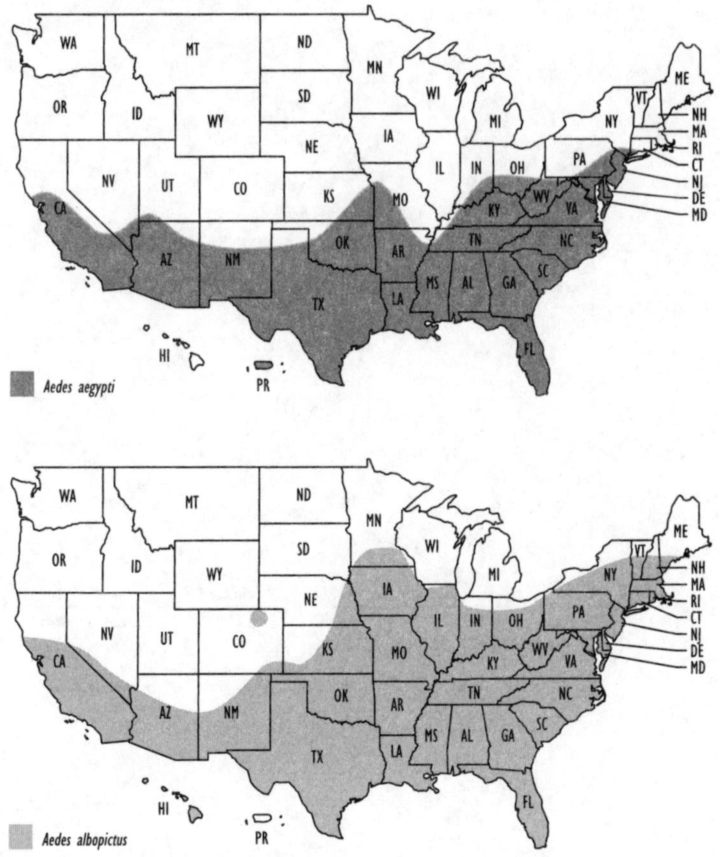

Estimated range of *Aedes albopictus* and *Aedes aegypti* in the United States, 2016. Redrawn from a map from the Centers for Disease Control and Prevention.

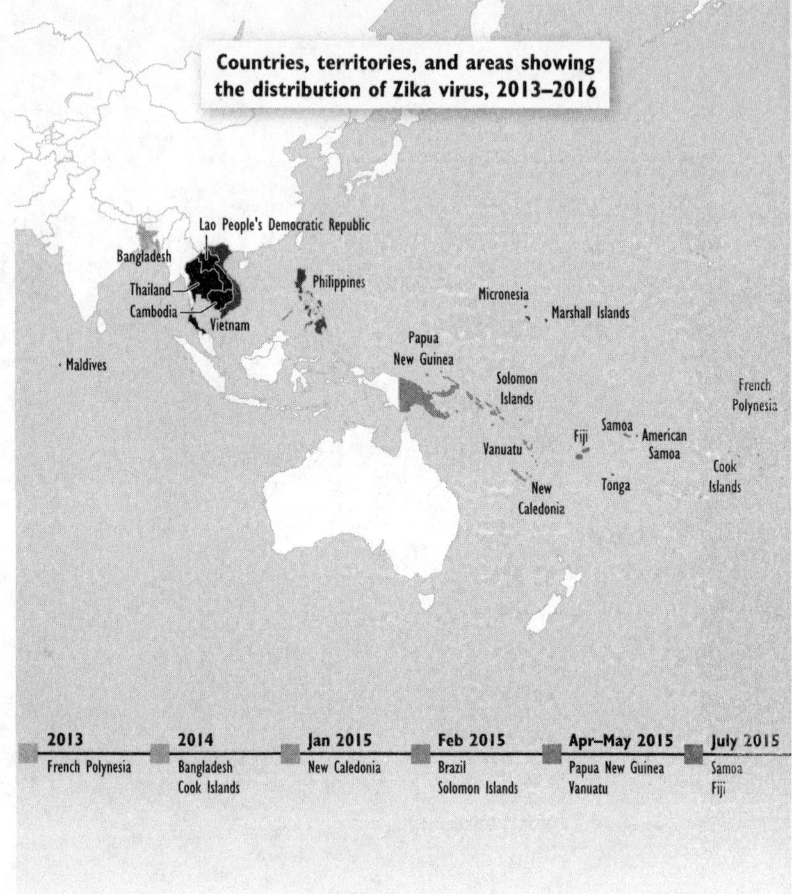

Countries, territories, and areas showing the distribution of Zika virus, 2013–2016. Redrawn from map printed in WHO, "Situation Report: Zika Virus, Microcephaly, and Guillain-Barré Syndrome," May 26, 2016, p. 4, http://www.who.int/emergencies/zika-virus/situation-report/en/. © 2016 by World Health Organization.

Haiti
Dominican Republic
Puerto Rico
United States Virgin Islands
Saint Barthelemy
Saint Martin/Sint Maarten
Guadeloupe
Barbados
Martinique
Saint Lucia
Saint Vincent and The Grenadines; Grenada
Bonaire, Netherlands
Trinidad and Tobago
Guyana
Suriname
French Guiana
Mexico
Cuba
Jamaica
Aruba
Curaçao
Belize
Guatemala
El Salvador
Honduras
Nicaragua
Costa Rica
Panama
Venezuela
Colombia
Ecuador
Peru
Brazil
Bolivia
Paraguay
Argentina
Cabo Verde
Oct 2015
Colombia
Cabo Verde
Nov 2015
El Salvador
Guatemala
Mexico
Paraguay
Suriname
Venezuela (Bolivarian Republic of)
Dec 2015
French Guiana
Honduras
Martinique
Panama
Puerto Rico
Jan 2016
Bolivia (Plurinational State of)
United States Virgin Islands
Dominican Republic
Costa Rica
Guadeloupe
Saint Martin
Nicaragua
Barbados
Maldives
Ecuador
Guyana
Jamaica
Curaçao
American Samoa
Haiti
Tonga
Feb 2016
Peru
Marshall Islands
Saint Vincent and The Grenadines
Sint Maarten
Trinidad and Tobago
Aruba
Bonaire, Netherlands
Mar–May 2016
Saint Barthelemy
Micronesia (Federated States of)
Vietnam
Belize
Saint Lucia
Cuba
Philippines
Argentina
Grenada

ZIKA

1

The *Doença Misteriosa*

In August 2015, something strange began happening in the maternity wards of Recife, a seaside city perched on the northeastern tip of Brazil where it juts out into the Atlantic.

"Doctors, pediatricians, neurologists, they started finding this thing we had never seen," said Dr. Celina M. Turchi, an infectious diseases specialist at the Oswaldo Cruz Foundation, Brazil's most famous scientific research institute.

"Children with normal faces up to the eyebrows, and then you have no foreheads," she continued. "The doctors were saying, 'Well, I saw four today,' and 'Oh, that's strange, because I saw two.' "

Some of the children seemed to breastfeed well and did not seem to be ill, she said.

Others cried and cried, in a weird, high-pitched wail, as if they were in constant pain and could not be comforted.

Some had seizures, one after the other, their tiny bodies wracked by spasms. If the seizures lasted long enough, they could disrupt the babies' breathing and heartbeat. Those babies often died after a few days.

Others seemed unable to flex their arms and legs, or their eyes jumped around erratically, not seeming to focus, perhaps not seeing anything.

Others did not react to noises and appeared to be deaf.

Others could not swallow. If they did not get intensive care and feeding through nose tubes, they, too, soon died.

As the horrified doctors compared notes, one thing stood out: many of the mothers mentioned that, months earlier, they had had the *doença misteriosa*—Portuguese for the "mystery disease"—that had first appeared nine months earlier in Recife, Salvador, Natal, and Fortaleza, the cities of Brazil's arid northeast.

Back then, the disease had not seemed a big deal. Everyone appeared to have the symptoms: an itchy pink rash; fever and chills; bloodshot eyes; headaches and joint pains.

Nonetheless, many people had gone to local clinics or emergency rooms because they were worried. The symptoms looked like the beginning of a few diseases known to kill people. They looked like early malaria or yellow fever. They looked most like the first symptoms of dengue, a disease Brazilians feared because of its unique power to deliver the double tap: The first time you got dengue, you were miserable. It was called "break-bone fever" because it felt as if someone had taken a sledgehammer to your arms, legs, and neck. Still, you usually recovered. It was the second bout that could kill you. A second infection—with a different one of dengue's four strains—was worse, and had a chance of turning into dengue hemorrhagic fever. As with Ebola, you bled from your nose and your mouth, under your skin, and from your organs. If you got hemorrhagic fever, you usually died.

And the number of cases was soaring. It was a powerful El Niño year, with both temperatures and rainfall much higher than normal. The mosquitoes were intense. Brazil had 1.6 million cases of dengue in 2015, nearly triple the 2014 caseload. Everyone knew someone who had had it, and nearly 1,000 would go on to die of it.

But the *doença misteriosa* had not been dengue. The rash and fever and headaches were unpleasant, but they did not get worse. No one died of it—unless the person was already seriously ill with something else. Almost no one was hospitalized. People went home from the hospital and told their families, "The doctors say it's not serious. They gave me pills for the headache, and there's no cure. But they say I'll get better in a few days. And, even if the kids get it, they'll be OK."

It was often misdiagnosed. Rumors attributed it to something in the water. Doctors thought it was an allergy, a parvovirus, or rubeola, or fifth disease, whose classic sign is called "slapped cheek" because it gives children bright red cheeks.

In May 2015, when the Cruz Foundation finally said it was an obscure African virus called Zika, the health minister in Brasilia sighed in relief. "Zika doesn't worry us," Dr. Arthur Chioro told reporters. "It's a benign disease."

There was no way to avoid the mosquitoes, so everyone got it. They groused, they joked, they empathized with each other, but it was just part of life. Things could be worse.

And then the disease seemed to start fading. Everyone had had it—and no one seemed to get it again. So probably everyone was immune.

In fact, the doctors said that. As far as they knew, Zika was like smallpox, chicken pox, or measles: once you'd had it, you never got it again. It hadn't been studied long enough for anyone to be sure it conferred lifelong immunity, but the protection seemed to be quite strong and long-lasting. It was not like dengue, which was worse if you got it again. It was not like malaria, which you could get year after year.

People stopped worrying.

But by then the disease was on the move. From its epicenter in northeast Brazil, it was moving west, into Colombia, Venezuela, and Suriname, and south, toward the megacities of São Paolo and Rio de Janeiro, which was struggling to get ready to host the 2016 Olympics. No alarms were raised. Mosquito control efforts, such as they were, continued apace. Compared with the other mosquito diseases around—malaria, yellow fever, dengue, chikungunya—this one was mild.

And then, about nine months after the disease had overrun the northeast, the babies began appearing. The babies with the tiny heads.

They gave the first notice that the *doença misteriosa* was by no means harmless.

2

The Origins of the Virus

ZIKA VIRUS WAS discovered in 1947, in the Zika Forest of Uganda, in a monkey.

There are hundreds of obscure viruses in the world—little curlicues of RNA with names like Spondweni, simian foamy, and o'nyong-nyong. A nasty one from the Four Corners area of the American Southwest was repeatedly renamed because the locals kept objecting to each moniker. It was first the Four Corners virus, then Muerto Canyon virus, then Convict Creek virus. It is now officially the Sin Nombre virus—Spanish for "no name."

Once in a while, one of the viruses leaps out of obscurity and into the headlines: Ebola. SARS. West Nile. Spanish flu. Swine flu. Bird flu.

But Zika virus is like no other. As Dr. Anne Schuchat, the principal deputy director of the Centers for Disease Control and Prevention (CDC), the world's premier disease-fighting agency, has said, "The more we learn about Zika, the scarier it gets."

It is the only mosquito-borne virus that routinely crosses the placenta to kill or cripple babies. Scientists do not know why or how it crosses the placenta when other mosquito-borne viruses like dengue, yellow fever, West Nile, and Japanese encephalitis almost never do.

It seems to be able to do so at any time in a pregnancy.

It is the only mosquito-borne virus that is also sexually transmitted.

And the mosquito that transmits it, *Aedes aegypti*, known as the yellow fever mosquito, is in 30 U.S. states, not 12, as was originally thought. A related mosquito that might transmit it, *Aedes albopictus*, known as the Asian tiger mosquito, is found in almost every state—its range in the hottest summers touches parts of Maine and Minnesota.

"Ziika"—the spelling was shortened—means "overgrown" in Luganda, one of the main languages of Uganda. The Zika Forest is no longer remote. It's on the highway between Kampala, the capital, and Entebbe, the country's main airport.

Quite a bit has been chopped down, so it is now less than one-tenth the size of Manhattan's Central Park. But in 1936, the Rockefeller Foundation established its Yellow Fever Research Institute in Entebbe, seven miles south of the forest. The forest was convenient and buggy; it bordered on a papyrus swamp.

Caged "sentinel monkeys" were suspended in six tall towers. The towers reached to treetop level, where the mosquito population was different from that of the forest floor. The monkeys were lowered daily to be checked, and their rectal temperatures taken and graphed.

Under 1940s-era medical ethics, it was perfectly acceptable to chain monkeys in trees to get sick or die. (Nowadays, all work with primates must be approved by ethics boards.)

It was not ethical in 1947 for scientists to use Africans as bait. That was progress. White farmers in some parts of colonial Africa protected their cattle from tsetse fly diseases by paying "fly boys." Tsetse flies hatch near rivers and are attracted to dark colors—including black skin. Young men—the fly boys—would stand shirtless in riverside brush, slapping dead every fly that landed on them. At day's end, they were paid a bounty per fly. The risk they took was that tsetses carry the parasite for sleeping sickness, a human disease that leads to a horrible death. It resembles rabies; victims may be driven mad, attack their own families with machetes, develop an unquenchable thirst but feel that the touch of water is burning them. Only in the end do they lapse into the coma that gives the disease its name, and die.

On April 19, 1947, in the Zika Forest, a monkey known simply as Rhesus 766 developed a fever of 104 degrees and was taken down from its platform and brought to the lab for a blood draw. It was an Asian monkey, not an African one. That presumably is the reason it got sick and the virus was discovered. Zika no doubt circulated in African monkeys for thousands of years, and they would have evolved resistance to it.

In those days, of course, it was not easy for scientists to figure out what a monkey had. There were no DNA-sequencing machines. The double-helix structure of DNA had not even been described yet. There were also no electron microscopes to let scientists see something as small as a virus. (Viruses are far tinier than bacteria, which are easy to view under a regular microscope.)

The scientists reported that they had found a "filterable, transmissible agent" in the monkey's blood. That meant they had spun down the blood to separate and remove the red cells, white cells, and platelets, and then had pushed the remaining clear serum through a ceramic filter with pores tiny enough to remove any parasites, like the ones that cause malaria, and all the bacteria. Then the serum would be injected into a healthy monkey. If that monkey fell ill with similar symptoms, then the first monkey's blood had contained a "filterable, transmissible agent"—likely a virus.

But then two far more complicated questions had to be answered: Was whatever made Rhesus 766 sick something new, or just one of dozens of other mosquito-borne viruses that caused similar symptoms? And how did one know that Rhesus 766 fell ill from something in a mosquito bite, and not from something it ate or touched, or from a biting fly or some other source?

Answering those questions took five years. The three scientists who did the work—Alexander J. Haddow and Stuart F. Kitchen of the Rockefeller Foundation and George W. A. Dick of the National Institute for Medical Research in London—did not publish their findings until 1952.

The search also consumed thousands of albino mice—"Swiss mice from Carworth Farms, New York," according to the original paper, published in the journal *Transactions of the Royal Society of Tropical Medicine and Hygiene*. The mice were all experimented upon when they were between 35 and 42 days old. Just breeding and feeding them, and keeping track of their ages, was a time-consuming job.

The first roadblock was that Zika virus didn't naturally make mice sick. When serum from a sick monkey was injected into their abdomens, nothing happened.

But if a tiny amount was injected directly into their brains, some became somewhat sick. The virus was "neurotropic," meaning it homed in on nerve cells, including brain cells.

So the scientists had to start "passaging" the virus through a series of mice. They took the brain of the first mouse to fall sick, made a slurry of it, diluted it with saline, centrifuged and filtered it, and injected some of that into another mouse brain, and then waited however long it took for that mouse to get sick. By repeating that process 17 times, they forced the virus to "adapt" to growing in mice. At the end, they had something they could reliably inject into mice with the knowledge that they would fall ill and probably die. It was no longer the pure "wild-type" Zika virus that in the forest was moved from monkey to monkey by mosquitoes. But it was close—and it worked in an animal model.

To make sure Rhesus 766 had caught a mosquito virus, they had to do a separate, parallel series of experiments. Other traps in the same platforms caught hundreds of mosquitoes. They had to be hand-separated by species for testing. Then batches of *Aedes africanus* mosquitoes were chilled to kill them, ground up, diluted, centrifuged, and filtered to produce a "supernate," an extract that the scientists hoped would contain enough virus from the guts and tiny salivary glands of some of the mosquitoes in the batch to make a monkey and/or a mouse sick.

One batch of 86 mosquitoes trapped on January 12, 1948—afterwards known as lot E/1/48—did the trick. It made mice somewhat ill.

After that, a long series of experiments ran in parallel to show that the filterable agent taken from the blood of Rhesus 766 and the filterable agent from lot E/1/48 were the same, and that they weren't any previously known virus.

That was done by testing the mystery virus against "convalescent serum," that is, against whatever magic component was in the blood of monkeys that the scientists had deliberately made sick by injecting them with the virus—and then had recovered.

Rhesus 766 hadn't died of Zika. It had gotten better, so the scientists knew that blood taken from it one month later had to contain whatever mysterious agent had "neutralized" the virus.

We now know that agent to be antibodies—tiny Y-shaped proteins that glom onto viruses, attaching all over their shells.

Viruses' outer shells—actually called envelopes—have spikes that fit, like keys into locks, onto receptors on the outsides of cells. Cold and flu viruses, for example, have spikes that perfectly fit the surface receptors on the cells lining human noses.

Viruses come in many shapes. The long, wiry Ebola virus is a filovirus, from *filum*, Latin for "thread." SARS, which is covered with spikes, is a coronavirus, from the Latin for "crown" or "halo." Zika is a flavivirus, and that family is instead named for its most famous member, yellow fever, which turns its victims yellow from jaundice. *Flavus* is Latin for "yellow."

There are more than 70 flaviviruses, including dengue, West Nile, and Japanese encephalitis; most are spread by mosquitoes or ticks. Under an electron microscope, they all look like little balls or spheres, but up close they turn out to be 20-sided polygons called icosahedrons. They resemble sinister Christmas ornaments. Inside each hollow ornament is the payload—a strand of RNA about 10,000 nucleotides long that it injects into the cell it invades.

The RNA turns itself into DNA and hijacks the internal machinery every cell uses to copy its own DNA and make new cells. Like commandos invading a town and converting its car factory into a bomb factory, the virus makes thousands of copies of itself. Eventually, the cell explodes, and the viruses are released to attack other cells, spreading the illness.

One part of the body's immune response is a set of white blood cells that engulf and "inspect" each new virus that enters the body. The cells measure the shape of its spikes and generate millions of new antibodies perfectly fitting that shape. When there are enough antibodies in the blood, and each virus's spikes are covered with matching antibodies, the viruses can't attach to new cell victims. The infection dies out. The host recovers. The antibodies stay in the blood for weeks, still passively on patrol.

Now the scientists had something from Rhesus 766 they could kill a mouse with, as well as something else that matched it, neutralized it, and would save the mouse.

They began asking fellow scientists to send them samples of other viruses and the antibodies that neutralized them.

Several Rockefeller Foundation laboratories obliged. So did Albert Sabin, a virus researcher in Cincinnati, later famous for his polio vaccine. Others samples came from the Wellcome Veterinary Research Station in Frant, England, and from the Virus Reference Laboratory in London.

They started testing them in mice.

If, for example, they infected a mouse with yellow fever, and the mystery antibody didn't save it from death, then the antibody wasn't to yellow fever. To cross-check that, they would infect another mouse with the mystery virus, and if the yellow fever antibody didn't save it, the mystery virus wasn't yellow fever.

It had to be done in many mice at several different strengths, because they weren't sure how much antibody was needed to neutralize how much virus.

And because mice sometimes just spontaneously died, it had to be done in batches of six mice, and the number of dead mice counted, to remove the element of chance. In theory, if they infected six mice with yellow fever, and gave them yellow fever antibody, all six would remain alive. If they obtained a mismatched antibody, all six would die. But another confirmation was seeing a relationship of dose and response. If a very diluted solution killed only one of six, but a strong solution killed all six, they must be on the right path.

When it was all over, they were able to say that their mystery virus wasn't yellow fever, dengue, West Nile, Eastern or Western or Japanese or St. Louis encephalitis, louping ill, Canfield B, Ilheus, lymphocytic choriomeningitis, Bunyamwera, Semliki Forest, Ntaya, or Bwamba fever.

It was "hitherto unrecorded," they said, and therefore a new discovery. They named it Zika.

But for the next 60 years, until 2007, it was barely heard of. In all that time, only 14 active human infections were described.

The first was described in 1952, by British health authorities investigating an outbreak of jaundice in the Afikpo Division of eastern Nigeria, which was then a British colony.

It was in "an African female aged 10 years." She was brought to a clinic because she had fever and headache. She was not jaundiced, but she had a fever of 100.8 degrees.

The two other patients in the study were men, aged 24 and 30. Both had antibodies to Zika but not the virus; only the girl had something in her blood that made mice sick.

By an abbreviated version of the mouse tests done in the first paper, it was shown that she did not have yellow fever, West Nile, Bunyamwera, Bwamba, Ntaya, Mengo, and so on—and that she did have Zika.

The author of the paper describing the first human Zika infection was Francis N. Macnamara, acting director of the Virus Research Institute of Yaba, Nigeria. A lot of top-notch research in tropical medicine during the colonial period was done by British, French, and Belgian scientists, much of it to keep African workforces alive and the troops of the colonizing power healthy. Dr. Macnamara's institute was the foundation for the Nigerian Institute of Medical Research, which is in the Yaba district of Lagos, the country's financial capital. Macnamara noted that the young girl's blood "contained numerous malaria parasites" but reassured readers that "in tropical Africa, infection with more than one pathogen is the rule rather than the exception" and that his tests were not confounded by the presence of malaria.

The 10-year-old female was reported to be "completely recovered six weeks later." So neither the Zika nor the malaria was fatal. That is not surprising; even today, kids in malarial regions of rural Africa who live past their fifth birthday have usually had malaria so many times that they are largely immune. It normally gives them just a debilitating fever.

Dr. Macnamara's paper was partially off base. He was investigating a big outbreak of jaundice, so its chief concern was whether or not Zika causes jaundice. (It generally doesn't—the poor girl was caught up by accident in an investigation of what was probably a completely different disease.)

But the paper contained a couple of very interesting asides.

One notes that the strain found in Nigeria "became adapted to mice more readily" than the original strain found in Uganda. Macnamara speculated that this was due to the fact that the Uganda strain was found in a forest, "whereas the Nigerian strain was probably well-adapted to man."

The paper also mentions, just in passing, a set of blood tests that the Yaba institute did on residents of the town of Uburu. Of the 84 residents tested, 50 had antibodies to Zika virus.

The first deliberate infection of a human with Zika was reported in 1956. It was in a human volunteer described as "a 34-year-old European male who was resident in Nigeria for a period of 4½ months prior to inoculation and had not contracted any known infection during that time." In other words, it was a new researcher at Dr. Macnamara's laboratory, William G. C. Bearcroft, who decided to infect himself.

After marking the spot on his left arm with an indelible pencil, he injected a "6th mouse brain passage material of the Eastern Nigeria strain of Zika virus (Macnamara, 1954) which had been preserved in sealed ampoules in the dessicated state at a temperature of –50 degrees C. for a period of 2 years."

He got a mild headache, a low fever—and no jaundice. He also let *Aedes aegypti* mosquitoes feed on him and then later on mice, hoping they would transmit the virus. Many of the mosquitoes died for unknown reasons, and none of the mice got the virus.

There is a long history in medicine of researchers testing things on themselves. Modern ethics boards frown on the practice, but some important discoveries have been made that way. In this case, Dr. Bearcroft didn't learn very much other than that Zika probably did not cause jaundice.

Jaundice—caused by the buildup of bilirubin, indicating liver damage—was important since it was the classic sign of yellow fever, a dangerous disease. Reporting that a patient has jaundice (from *jaune*, French for "yellow") literally means that his skin and the whites of his eyes have turned yellowish.

In 1964, another researcher, this time a 28-year-old European male who had been in Africa only two and a half months, argued that the girl and Bearcroft had probably never had Zika, but rather Spondweni, a related virus. He claimed *he* was the first person to be able to scientifically describe the symptoms of Zika, because he had just had it. He was David I. H. Simpson. Simpson was a student of Dr. George W. A. Dick's when he taught microbiology at the Royal Victoria Hospital in Belfast, Northern Ireland. Dick encouraged him to work abroad, and he moved to the East African Virus Research Institute in Entebbe, Uganda (which was a later name of the Rockefeller Foundation's Yellow Fever Institute). Simpson said he contracted Zika in the course of his work with the virus at the Entebbe lab, which he had just joined.

Interestingly, he was the only one to develop the most characteristic sign of Zika: "a diffuse pink maculopapular rash" on his torso, face, and upper arms that lasted for five days, finally spreading all over his body. He called the disease "mild."

After that flurry of interest, the virus appears only sporadically in medical history before 2007. Between 1960 and 1983, cases were detected in the Central African Republic, Gabon, Senegal, Ivory Coast, Cameroon, and Sierra Leone.

At some point—perhaps in the 1960s, perhaps earlier—it moved to Asia. It was identified in Malaysia in 1969 and in Pakistan and Indonesia as early as 1977.

Ultimately, a strain began to cross the Pacific. (Later genetic sequencing determined that it most closely resembled a 2010 sample from Cambodia, but so little sampling was done back then that there is no guarantee it started in that country. There is much more air traffic between the South Pacific and other places, like Indonesia.)

Why wasn't it studied more?

And why didn't it cause outbreaks of microcephaly during that time?

The first question has several answers.

Zika wasn't studied because it was rarely even diagnosed. Its symptoms resembled those of other, more serious diseases, notably dengue. Those diseases were often circulating in the same country, so a doctor seeing a rash and fever would probably shrug and say, "It's dengue, but mild. You're lucky." There was no point in sending a sample away—and nowhere to send it to. No lab routinely did Zika tests. Modern labs need "primers" for their PCR machines, a thermal cycler for amplifying DNA. The primers are short sequences of half the DNA "ladder" that match the other halves being run through the machine. For routine tests, primers are for sale in many forms. For extremely obscure viruses, a lab would have to create its own.

Moreover, Zika was never considered important. Everyone thought it didn't kill people or even hospitalize them. The scant medical literature on it described it as mild in humans. It also didn't harm any valuable farm animals like chickens, cattle, pigs, or even camels.

In the hunt for research funding, a virologist specializing in Zika would struggle to get grants, while those studying bird flus would see the dollars roll in because of the threat to the poultry industry and, later, the possibility that avian flu would kill millions of people. (The panic of a decade ago about avian flu is over. The threat is not.) That is a serious disincentive, and there are many orphan viruses that are known about, but not studied.

And as the early papers showed, Zika was hard to study because there was no reliable animal model. Small, docile, fast-reproducing creatures like mice, rats, gerbils, and rabbits are ideal, but they don't always cooperate. The best model for human flu, oddly enough, turns out to be ferrets; human flus reliably make them lose weight, become lethargic, and sometimes die. But live ferrets—big furry weasels—are fast, fierce, and famous for a vicious bite.

Monkeys get Zika, but don't reliably fall ill from it. And monkeys have enormous drawbacks as animal models: they are expensive, they take lots of care and feeding, they bite, they throw feces, and they are smart enough to notice a missing or open lock and escape. They are also adored by animal-rights activists, who may conduct raids to free them. Moreover, monkeys caught in the wild may have unpredictable diseases that infect other primates, including humans. In 1989, a batch of crab-eating macaques shipped from the Philippines to Reston, Virginia, turned out to have a relative of the Ebola virus that jumped from monkey to monkey in the animal house and was also caught by a handler who cut himself working with them. Luckily for everyone in the lab—and possibly for everyone in the United States—that viral relative was not lethal to humans. The outbreak was ended by killing all the monkeys, sterilizing the building, and then demolishing it. It is now known as Reston virus.

Haddow, Kitchen, and Dick had developed a mouse model through cumbersome "serial passage" through many mice. "Passaging" is a common technique in virology. For example, the "spines" of human flu vaccines were made by passaging human flu viruses through many generations of fetal chicks. When viruses are adapted to growing in chicks, they no longer reproduce easily in humans—and they can be grown in chicken eggs. Every year, millions of fertilized chicken eggs are used to grow flu vaccine. One crucial question for vaccine makers each year is whether they will have enough roosters to fertilize the eggs. Aging but still spry cocks destined to end up on supermarket shelves as ground-poultry products get temporary reprieves each year because they are on call to perform a vital task for the vaccine industry.

But a virus that emerges at the end of a long series of mouse passages and attacks mouse nerve cells is no longer exactly the same as the virus that infects monkeys and humans. Scientists can only hope that any discoveries they make—any drugs that kill it off in those mice, for example—will work in humans, too.

Only this year (2016) did scientists come up with easy mouse models for Zika. Nowadays there are dozens of strains of genetically altered mice for sale—mice that routinely develop the symptoms of Parkinson's, multiple sclerosis, or Alzheimer's, for instance. They are variants of the first "knockout mice." Different genes in their DNA have been "knocked out," or silenced.

In March 2016, researchers at the University of Texas Medical Branch in Galveston announced that they had found a set of off-the-shelf mice known as AG129, which lack the genes to mount an interferon-based immune reaction, would succumb to Zika. Notably, the virus killed fetal mice but not adults, which paralleled its effect on humans. It was found concentrated in their brains, as it was in human fetuses. It also concentrated in their testes, as it was suspected to do in adult men. That made it a good model, although others were likely to be found, the researchers admitted.

As to why it circulated for decades without causing microcephaly, we can only guess.

In Africa, the answer is easy. It no doubt circulated there for centuries. The blood tests from Uburu showed that of 84 residents tested, 60 percent had had the virus. Children in most of Africa get thousands of mosquito bites as they grow up. If only a few of those bites had Zika in them instead of malaria, they would get it, recover, and be immune. If most girls, like the 10-year-old in the Afikpo Division, became immune before their child-bearing years began, they would never pass it to their babies.

Why it never caused microcephaly in Asia is still a puzzle.

There's no certainty about how long it circulated there. It also remains unknown whether there were other factors, like a previous bout of dengue, that predispose some women to more dangerous infections.

Another possible answer is that it *did* cause microcephaly—but that no one noticed. There have always been microcephalic children in Asia, as there are everywhere else, because the condition has many causes. Some degree of it occurs in between 1 in 5,000 and 1 in 10,000 births. Mothers can get infected during pregnancy for the first time with *Toxoplasma gondii* (a bacterium found in cat feces, which is why pregnant women are told to avoid cat litter boxes), with cytomegalovirus, herpes, or syphilis. It can also result from fetal alcohol syndrome, from drug abuse, from exposure to some industrial or agricultural poisons, or from severe malnourishment in the mother. And it can be caused by genes, like those that cause Down syndrome.

A likely explanation may be that there *were* clusters over the decades, but they were blamed on rubella—German measles. In unvaccinated populations, rubella epidemics wax and wane. The virus blows through a population, infecting everyone, causing damage but creating herd immunity, and then disappears for a decade or more. It can't return until enough new victims have been born to sustain a new epidemic. That's why, in the prevaccine era, highly infectious diseases were called childhood diseases. Most teenagers and adults had already had them. But sometimes the gaps between epidemics were long enough that many young women entered their child-bearing years unprotected, as occurred in the United States in 1964, a year that saw a lot of birth defects.

Until the huge efforts to vaccinate the world's poorest children emerged in the last 15 years—thanks largely to the Bill and Melinda Gates Foundation and the generosity of American and European taxpayers—epidemics of childhood diseases like rubella and measles were far more routine in poor countries than they are now.

Zika can hit anyone, but it is more likely to hit poor people, who live in slums with open gutters and piles of rain-collecting garbage where mosquitos breed. Poor people are also more likely to be exposed to other causes of microcephaly: not being vaccinated against rubella, living where feral cats roam, being poisoned by industrial chemicals in shantytowns that spring up near factories, suffering from severe malnourishment.

In addition, poor people are less likely to give birth in hospitals. Home birth is a strong tradition in much of Asia. Even today, in India, Bangladesh, Pakistan, and elsewhere, many women are under family pressures to give birth at home with a traditional attendant. Going to a hospital may be seen by their grandparents as bowing to a foreign, Western medical tradition. And it costs money. Even in "free" public hospitals, doctors and nurses live on meager salaries, and their pharmacies are often empty. It's not uncommon in poor countries to see a row of tiny pharmacy stalls outside the gates of big hospitals. They have the drugs that the hospital does not. A nurse knows what a patient needs, goes out to buy it, and charges the patient. So a young girl in Bangladesh may be pressed to both stand up for her culture and save the family money by giving birth on a floor mat.

Microcephalic babies born at home are never counted. Without intensive care, some die quickly. Some that do live may just be hidden in the house out of shame—fear that they mean someone in the family angered the gods, for example. Brazil's cluster was noticed because it took place in hospital wards. South America's largest country, Brazil still has an emerging economy. It has some first-class hospitals, and even the poorest, most traditional families have heard that the outcomes for mothers and babies are better there than they are giving birth on the floor at home. So they go.

At some point—no one is sure exactly when or where—Zika broke out of Asia. It would spend the next few years leapfrogging from island to island across the South Pacific like the Marines in World War II, but in reverse.

3

On the Move

THE FIRST TIME Zika was noticed outside of Africa and Asia was in 2007, when Thane Hancock, a family physician working for the Yap Department of Health Services, sent an email to the CDC in Atlanta asking for help. Yap is one of the Caroline Islands in the western Pacific, and about 500 Yap islanders, Hancock said, had come down with something that resembled mild dengue, but didn't come up positive on the dengue test kits the island had on hand.

Yap had a mere 7,000 inhabitants, so 500 cases constituted a big outbreak.

Yap is part of the Federated States of Micronesia. In World War II, a Japanese bomber group and about 6,000 troops were based on it, and the U.S. Navy bombed it repeatedly. With Japan's surrender, the United States seized the islands. They became independent in 1986, but signed a Compact of

Free Association with the United States. They are not a territory like Puerto Rico or Guam, but still loosely attached, so when they needed help, they turned to the CDC.

The email arrived in late May and was forwarded to the Epidemic Intelligence Service. The EIS is the CDC's elite division of disease detectives, and competition to get into each year's class of 75 trainees is stiff. Its symbol is a globe with a shoe sole superimposed on it, and the sole is worn through, like a detective's who will stop at nothing. The service investigates about 100 outbreaks a year—everything from *E. coli* killing fast-food customers in the Midwest to rashes on an island half a world away.

On June 13, Lieutenant Colonel (Dr.) Mark Duffy, an Air Force epidemiologist assigned to the division of vector-borne infectious diseases, and Dr. Tai-Ho Chen, a medical officer in the EIS Field Assignments branch, arrived on Yap. They immediately started seeing patients in the island's five clinics and sent samples to the agency's arbovirus laboratory in Fort Collins, Colorado.

Dengue kept looking like the most likely explanation, Dr. Duffy said later, until the results came back on June 22. It was something new for the Pacific, and something that had never been seen causing a big outbreak before: Zika virus.

That was eye-opening. The scant existing literature described it as an African virus. Yap Island is a *long* way from Africa, by any route.

The team expanded to eight members—seven Americans and one from France's Pasteur Institute. They spent the next six weeks there.

"We worked through hot, humid weather punctuated by daily drenching rains," Dr. Chen later wrote for the *EIS Bulletin*. "We completed household survey activities despite heavy rain and winds of Tropical Storm Man-Yi (later a Category 4 typhoon when it hit Japan). Memorable events include the team eating a reef fish caught by Mark, drinking home-brewed beer, and listening to our French entomologist playing the ukulele and singing 'Drunken Sailor.'"

Like the British authorities in Nigeria, the CDC did a serosurvey. Serosurveys are like opinion polls: you pick a representative sample of the population, meet the selected people, and dig deeply into the data they give you. But after asking them lots of questions, you check their answers by looking in their blood.

The team went to 170 households, inquiring about a family's recent symptoms, taking blood, making notes about the household environment (standing water, screens, etc.), and collecting mosquitoes. It also went to all the island's hospitals and clinics, pulling records and interviewing doctors.

The epidemic, they realized, had peaked in May, just as Dr. Hancock was asking for help. In all, it lasted only five months. But by screening the blood for antibodies and looking at who had reported symptoms, they became the first scientists to figure out the dynamics of a Zika epidemic in a "naïve" population—one in which no one was immune.

They calculated that 73 percent of the island got the disease in that five-month window. By August, it was over. Cases disappeared, and there has not been one on Yap since—presumably because herd immunity is so high. Almost everyone is immune, so even if one naïve person brought it back, there wouldn't be enough susceptibles around to let a new outbreak start.

Four out of five who got the disease never knew it. They showed no symptoms.

And it had been universally mild. No one had been seriously ill. No one had died.

"Our health care system is doing fine," Dr. Hancock told a Reuters reporter who telephoned from Hong Kong. "We haven't been overloaded by heaps of patients coming in."

The virus's next appearance would be a bit different.

Zika wouldn't be heard from again for another six years, and it would be on another dot in the Pacific 5,000 miles to the east: Tahiti, the main island of French Polynesia.

French Polynesia consists of 118 islands scattered over an area 10 times the size of France. The whole country has a population of only 270,000, with about 75 percent of it living on Tahiti or neighboring Moorea.

Polynesia is much more closely connected to France than Micronesia is to the United States. As an "overseas collectivity," it has some autonomy, but it sends deputies to the National Assembly in Paris and is patrolled by French troops and gendarmes. It benefits from the connection in various ways. One is that it has had a very impressive medical surveillance network in place since 2009, with 50 sentinel sites on 25 different islands—a mix of public and private clinics. The doctors at those sites saw almost a quarter of the population, and they were expected to file weekly reports to

Papeete, the capital. At the top of the chain was the national hospital and the Louis Malardé Institute, which had connections to the Pasteur Institute in Paris, one of the world's top medical research institutions.

On October 7, 2013, the first alert went out: clinics on several islands were reporting an "eruption" of patients with fevers, rashes, bloodshot eyes, and painful and swollen joints.

Blood samples began coming into the Malardé Institute. At first, a new outbreak of dengue was suspected. There are four strains of dengue, and types 1 and 3 had both been seen in the islands since February. But Van-Mai Cao-Lormeau, the head of the institute's laboratory, was doubtful.

"Tahiti is a small island," she told NPR News later. "So in the lab we had relatives, family, and friends who were getting sick who we knew had already had dengue several times."

Even a first bout of dengue can be painful, and a new infection with a different type is usually much worse. But these cases were consistently mild. Dr. Cao-Lormeau's lab was unusually well-prepared. Because it did regular dengue testing as a courtesy for Micronesia, it not only knew about the Yap investigation but had the CDC's Zika-testing protocols.

The first household tested for Zika contained a 53-year-old Tahiti resident, her 52-year-old husband, and 42-year-old son-in-law. They were negative for dengue, chikungunya, and West Nile, and "inconclusive" for Zika. But soon afterwards, a 57-year-old man came up positive. After that, more than half of the next 700 samples tested came up positive, so the lab finally stopped bothering.

It was official: the world's second major Zika outbreak was on.

That confirmation, issued on October 30, reported a total of 600 suspected and confirmed cases from multiple islands, suggesting that the virus had spread quietly for weeks beforc even French Polynesia's impressive surveillance network picked it up. Within two months, it had reached all 76 inhabited islands across the five sweeping archipelagoes that make up the territory.

Dr. Didier Musso, chief of the institute's emerging diseases unit, told the local government about it and asked Paris for help. But the response was initially tepid. After all, the CDC had described Zika as mild.

Then, in early November 2013, something alarming happened.

Patients began arriving at emergency rooms in varying degrees of paralysis. Most reported having had Zika symptoms in the last 15 days.

The first was a woman in her 40s, with partial paralysis of her arms, legs, and face. She was treated with immunoglobulin and, within two weeks, went home. Her blood was sent to a military hospital in France, which found that, besides Zika, she had antibodies to all four types of dengue—some old infections, some new.

Dr. Sandrine Mons, head of the intensive care unit at the national hospital, recognized the paralysis cases as Guillain-Barré syndrome. It was not unheard of—the country had 3 to 5 cases every year, although there had been a bump of 10 cases in 2010, which had been a big year for dengue. But all of a sudden, there were dozens of victims.

"Up till then, everyone thought this was a benign disease," Dr. Mons later told *Le Figaro*. "But as the Guillain-Barré cases kept going up, people began to be afraid. We ultimately had 42 cases, 16 of them in intensive care. The ones who were completely quadriplegic, with their breathing muscles paralyzed, had to be on a ventilator and artificial feeding for one to two months." Even some of the less serious patients had brain inflammation, persistent partial paralysis of the face or one side of the body, and vision problems. One developed a heart rhythm problem.

Guillain-Barré is usually temporary but always frightening. Victims have typically recovered from an earlier flu, stomach virus, or bacterial infection and think they are out of the woods. Then they start to sense that something is wrong—often just a tingling in the hands and feet and a sense of malaise. But as it progresses up the limbs, it feels as if they were turning to stone. It can stop—or not. If it reaches the diaphragm and chest walls, a patient who isn't ventilated immediately will die, wide awake and terrified, staring at the ceiling.

It's an autoimmune reaction to the earlier infection, in which the immune system produces antibodies that attack the body's own peripheral nerve cells. There is no cure. There are two types of treatment that can speed recovery. If a hospital has a plasmapheresis machine—or, during an outbreak, enough of them—"plasma exchange" is done. The machine draws blood from one vein, separates out the red and white blood cells and the platelets and returns them to the patient through another vein. The clear plasma, which contains the dangerous antibodies, is discarded. The body is forced to make more plasma, and, if the autoimmune reaction has died down, it won't have the antibodies. The alternative is immunoglobulin treatment: patients get large doses of plasma from healthy blood donors. Their antibodies somehow block or counteract the dangerous ones. Immunoglobulin therapy has its risks, including the transmitting of viruses. But if no plasmapheresis machine is available, there may be no choice.

Guillain-Barré usually fades away, but it can take many months. Some victims never fully recover muscle tone. Some live with constant pain.

Four cases of "immune thrombocytopenic purpura"—a condition related to Guillain-Barré—were noted. Nothing much was made of them at the time, but that condition would come up later when the epidemic reached American soil.

Papeete's rehabilitation center, used to housing only a few patients, suddenly had 18 with serious neurological problems, and it struggled to cope.

The worst-off was Larry Ly, a big, broad-shouldered, 42-year-old soccer-playing maintenance technician. On December 3, he drove his daughter to school, came home feeling bad, and lay down. Within two minutes, starting from his feet, he became totally paralyzed, he told STAT news later. "I was lucky it didn't happen when I was driving," he said. He struggled to breathe in the ambulance, and a hole had to be cut in his neck to intubate him.

He was in intensive care for eight months and was still in rehabilitation earlier this year to recover from surgery to free up an arm that prolonged paralysis had frozen in place.

As fear of the disease rose, the government stepped up mosquito spraying. Then the rumor spread that the insecticide, deltamethrin, and not some mystery virus, was responsible for the paralysis. Some mayors openly refused to spray their towns. French Polynesia's health minister, Beatrice Chansin, made a show of visiting paralyzed patients in the rehabilitation center and held a press conference at their bedsides to say spraying was crucial and deltamethrin was considered safe by the World Health Organization (WHO). Finally, the French high commissioner stepped in, reminding the mayors that, under French territorial law, if they failed to take precautions against "fires, floods, dike breaches, landslides, avalanches, or *epidemics of contagious disease*," they were not protected by their official status and could be held personally liable for the medical costs of their town's victims.

By April 2014, when the epidemic was officially over, the sentinel sites had reported 8,750 patients seeking care. The health department's epidemiologist, Dr. Henri-Pierre Mallet, calculated that 32,000 people, or about 12 percent of the country's population, had Zika symptoms severe enough to warrant a visit to a doctor. He concluded that the virus had reached 66 percent of the population.

The outbreak had been well underway by October 2013, when it was spotted. It peaked in December and had fallen to near zero by April. Not a case has been reported since August 2014.

The Guillain-Barré "attack rate" was judged to be 1 for every 4,200 Zika infections. That is about 25 times the normal background rate for the world, which is 1 for every 100,000 people per year.

The typical Guillain-Barré victim was a male over the age of 40.

None of the early reports mentioned microcephaly, or any particular problems for babies at all.

In an interview with the French news magazine *Le Point* in 2016, Dr. Musso was blunt and bitter. When he had asked for help, he said, the South Pacific Commission and the WHO had sent experts, but the government in Paris had largely ignored him. What help he did get came from the Pasteur Institute and a military hospital in Marseille, he said.

"We toughed it out alone to isolate the virus, develop diagnostic tests, manage the patients, and face the first serious unexpected complications," he said. "When you live at the far end of the world, you learn to cope."

In 2015, when France's High Council on Public Health had held a meeting of experts to issue Zika recommendations, he wasn't invited, he said. "Frankly," he commented, "the high authorities here never ask the opinion of people who've actually lived through the problems."

As a result, he remarked, the recommendations, which applied to many French islands, including those in the Caribbean, yet to be in the virus's path, were naïve: the council recommended testing only symptomatic pregnant women. They should have recommended testing all of them, he said, since 80 percent will have no symptoms but could still suffer.

From Tahiti, the virus spread quickly to other Pacific island nations. New Caledonia, another set of French islands off the coast of Australia, was first. On March 5, 2014, Easter Island, home of the giant stone heads and also called Rapa Nui, reported a case in an 11-year-old boy who had never been off the island. Rapa Nui is Chilean territory but ethnically Polynesian, and a month earlier it had hosted the Tapati festival, the largest cultural event in the Pacific, which many French Polynesians attended. On March 11, the Cook Islands, next to Polynesia, confirmed a case.

But that was paid attention to only later. Outbreaks on remote islands rarely make headlines even in the unusual case in which they are truly scientifically investigated, as the ones in Yap and French Polynesia were. The virus effectively "disappeared" again.

One of the epidemic's great unanswered questions is how it made the leap from that scattered medley of oceanic nations to northeast Brazil, which juts far out into the Atlantic, has no overt cultural ties to Polynesia, and is geographically much closer to Africa.

On the map, the closest suspect is Easter Island. It's in the Western Hemisphere, directly due south of Salt Lake City. But its air connections are back to the Pacific and to Chile, which has not had a case yet and is not expected to, because its climate is too cold for *Aedes aegypti* mosquitoes.

Brazil didn't even realize it had Zika until May 2015.

The first impulse of many Brazilians was to blame the soccer World Cup championship, which was held in June and July 2014. Stadiums in Recife, Natal, and Salvador, northeastern cities eventually at the epicenter, had all played host to games. But although the World Cup draws tourists from all over the world, no South Pacific nation had played in it.

Dr. Musso then published a letter suggesting that a more likely explanation was that it had arrived during the Va'a World Sprints, a set of outrigger canoe races held in Rio de Janeiro a month later, in August 2014. About 2,000 paddlers arrived for it, including teams from French Polynesia, New Caledonia, the Cook Islands, and Easter Island.

But in March 2016, genetic sequencing of the virus let scientists construct a "molecular clock" of how fast it had mutated as it spread. By "winding back the clock," they estimated that it had been in Brazil since mid-to-late 2013.

Another discovery gave credence to that idea.

In April, researchers at the University of Florida went back and looked at a big batch of blood samples from an outbreak of chikungunya in Haiti. The samples were from school clinics, and the blood of three students, aged 6 to 14, tested positive for Zika. The researchers checked the dates—they had all been collected in December 2014, which meant the virus was also in Haiti well before it was identified in Brazil.

That didn't mean it was there first. It may have circulated under the radar in Brazil, Haiti, and perhaps elsewhere, for months before some unusual set of circumstances produced an explosion in northeast Brazil.

Now the prevailing theory is that it was introduced during the FIFA Confederations Cup, a prelude to the World Cup. It too was played in Brazil, but a year earlier, in June 2013. It included a team from Tahiti, which played one game in Recife. That theory is a bit of a stretch, for it would be four months before Tahiti's outbreak was detected, and a year and a half until Recife's was. But viruses are sly.

This was not the first or even second time that Zika victims had blamed their misery on soccer. When the virus swept French Polynesia, rumormongers pointed fingers at the World Cup of Beach Soccer, which had been held in Tahiti in September, just before their outbreak was detected. One of the 16 teams in it was from Senegal, they said, and Zika was an African virus, wasn't it? But the Senegalese were innocent, because genetic testing done later showed that the Polynesian outbreak was virtually identical to the one on Yap, and descended from the Asian lineage.

4

The World Hears

The world heard about the mystery virus when it leapt out of Brazil in headlines above pictures of grieving mothers holding babies with heads that didn't look right.

They looked like Cabbage Patch Kids or Trollz dolls—all chubby cheeks and big eyes, but with dark hair sprouting too closely behind their foreheads. They looked proportional, but somehow out of proportion, and it took the viewer a second to realize that what was wrong was that normal babies' heads look too big for their bodies. These babies looked more like old men with wrinkled brows.

But that was just cosmetics. Babies often look odd—scrunched or wizened or yellow or cross-eyed, or even born with elongated or oddly shaped heads—and yet they can be perfectly healthy. The struggle through the birth canal can be hard on an infant's soft plasticity.

The real and terrible consequence could be seen on CT scans, MRIs, and ultrasounds. Those tiny heads contained shrunken brains. Sometimes just the frontal lobes—the seat of decision-making, of speech, of intelligence, of humor—were atrophied, showing abnormally large dark ventricles, the hollow internal spaces that are supposed to appear smaller and smaller as the brain grows. Sometimes all that was left was the bulb above the brain stem, where the most basic functions, like breathing and digestion, reside. Around it would be blank space filled with cerebrospinal fluid. Usually the skull had not completely collapsed, but neither had it been pushed out to its full size by the growing brain. And the brain would be smooth, looking more like a small liver, with none of the deep folds and fissures—the sulci and gyri—that every growing brain should develop as it folds in upon itself to pack more thinking power into a small space.

That smooth-brained baby might be more than comatose; maybe it could breathe, could blink, could digest, could live. But maybe that baby could not chew food, or see the spoon or the breast coming toward its mouth. Certainly it would never walk, probably would never crawl, or maybe would never do more than roll from side to side, unable to control its contorted arms and legs enough to even turn over.

Hospital hallways, doctors remembered in Brazil, were lined with mothers who resembled ghosts. They were in shock: mute, expressionless, bleak. Some were just teenagers. Some had ridden buses for hours and were too poor to buy food as the hours waiting to be seen stretched on. And there were so many of them. One doctor from southern Brazil, where there was no problem, recalled visiting a friend's hospital in Salvador, not at all expecting what he found: 25 babies with microcephaly, all born in the previous 10 days. One mother looked up from her son's face to ask, "Doctor? His head is going to grow, right?"

Those mother-and-baby pictures—normally records of happy occasions, now a series of postcards from hell—became the signature of Zika.

All over the world, pregnant women began to worry. So did everyone, man or woman, who hoped one day to have a child.

As well they might. Right now, at least 298 million people in the Americas live in areas "conducive to Zika transmission," according to a recent study. Which is a conservative count, because, if you count everyone who lives between northern Argentina and southern Tennessee—roughly the range of the *Aedes aegypti* mosquito—you get over 400 million.

Over the next year, according to that conservative study, more than 5 million babies are due to be born.

How much damage Zika will ultimately do is not yet knowable. The aggressive spread outward from Brazil's northeast began only in 2015, and most of the Western Hemisphere, including the United States, has not yet lived through even one full hot season with it.

What could happen if it spreads widely across Africa and Asia is a whole different level of disaster. About 130 million babies are born each year around the world.

Zika has been on those continents for decades, and many

Africans and Asians may be immune to it. On the other hand, the African and Asian strains are different from each other; the Asian one has several substrains, and viruses constantly mutate. The flu virus mutates so fast that the vaccines against it must be reformulated each year. The Zika virus is not that mutable, but it may have shifted enough that immunity to the old strains does not confer protection against the new one.

One aspect is reassuring: more than 99 percent of all cases are mild. Most adults, teenagers, and even toddlers who get it appear to come though unharmed. So do most pregnant women—they themselves, that is.

The great threat is to unborn children. How great is not known as of this writing. French Polynesia's experience suggested that mothers who had Zika while pregnant had a 1-in-100 chance of having a deformed child. A small study in Brazil suggested it was closer to 1 in 3. More research is being done.

After that, the greatest threat appears to be autoimmune reactions, the best-known of which is Guillain-Barré. As of this writing, it is thought to occur during Zika epidemics at 20 to 25 times its normal rate—that is, once in every 4,000 to 5,000 infections.

A very small number of people with other complicating illnesses, like sickle-cell anemia, have died while infected with Zika. But it is not believed that Zika inevitably hurts everyone with comorbidities. The sickle-cell trait comes from Africa—where it is a genetic defense against malaria—and is common in Brazil and the Caribbean, where many are descended from African slaves. But, as of this writing, deaths from it that are clearly related to Zika are very rare.

Also as of this writing, it is not thought that Zika particularly harms people whose immune system is suppressed, such as those with HIV, those taking antirejection drugs for organ transplants, or those whose bone marrow has been temporarily ablated to fight leukemia.

But the threat to babies is enough. The tiny virus, delivered by a mosquito that can be squashed with a finger, is rerouting cruise ships and Boeing 737s. It is canceling destination weddings and family vacations. It is threatening the 2016 Olympics, and has further shaken Brazil's already shaky government. Failures of other presidents to fight it aggressively enough may yet topple other leaders.

For many people—certainly many Americans—the scare may be brief: a vacation canceled, a business trip replaced by a phone call. For some, living in tropical climates, it will mean months of worry: Worry that each mosquito might be the dangerous one. Worry that they have a silent infection. For women who are pregnant, that worry might be sheer terror: having to ask themselves every day for nine months, "Is my baby all right? Was it my fault? Did I do everything I could to protect it?"

For more than 1,400 women in Brazil and elsewhere in the hemisphere, that terror has already arrived. They know their babies are not all right. That if they survive, they will need a lifetime of care, will need watching night and day. Careers will be dropped, houses will be sold, bank accounts will be drained; in the United States, the cost of such care is estimated at $10 million per child. They know the guilt and exhaustion and anger of having a handicapped child and may fear that it will tear their family apart. Overwhelmed husbands abandon overwhelmed wives, resentful siblings will rebel.

And a mother's worry does not end even on her deathbed: she may die wondering who will take care of the child for the rest of his or her life. Will those family caretakers have the money? Will they have the patience? Will they have the strength? And will they not hate her memory for leaving them the burden?

5

My First Brush

I FIRST HEARD the word "Zika" in September 2015, when a media rep for the University of Texas Medical Branch (UTMB) emailed me asking whether I wanted to interview Dr. Scott Weaver, scientific director of the school's Galveston National Laboratory, about chikungunya, a mosquito-borne virus. I knew the disease; its name comes from the Makondo language spoken in Tanzania and Mozambique and means "bending-up fever," for the way its victims twist miserably in pain. It doesn't normally kill, but the pain it causes can last for months. It was invaliding many Latin Americans and making travel riskier for American tourists.

I replied apologetically that I knew it was high time I did a big story on chikungunya, but that I was too busy right then.

It was an honest excuse. As the *New York Times*'s main

global health reporter, I'm supposed to track the world's vital signs and cover every pestilence and plague that comes down the pike, so among my worries are AIDS, tuberculosis, malaria, Ebola, avian flu, swine flu, seasonal flu, SARS, MERS, polio, Guinea worm, diphtheria, pertussis, tetanus, measles, mumps, rubella, rotavirus, norovirus, respiratory syncytial virus, Hib, smallpox, cholera, the Black Death, Lyme disease, West Nile virus, yellow fever, Rocky Mountain spotted tick fever, erlichosis, babesiosis, cutaneous and visceral leishmaniasis, syphilis, gonorrhea, chlamydia, human papillomavirus, anthrax, ricin, cryptosporidium, Chagas, Buruli ulcers, Lassa fever, mycetoma, and the common cold—which is caused by nearly 100 different viruses. I may have forgotten some.

Also, because distrust in science among Americans is powerful, I cover some controversial diseases and persistent myths like Morgellons disease, delusional parasitosis, chronic Lyme, and the notion that vaccines cause autism.

The UTMB media guy tried again: Dr. Weaver was also an expert in something new that was "similar to West Nile virus and could make its way to the United States in the next few years." It was called Zika virus. Would I like an interview about that?

Just the name had a certain zing to it. (I once interviewed a pharmaceutical executive who believed that putting the zee sound in the names of new medicines inspired a soothing credibility and sold more pills: Prozac, Zoloft, Xanax, and Zyrtec, for example. The vee sound, he thought, conveyed virility—hence, Viagra and Levitra. Their new rival,

Cialis, pronounced "see Alice," he thought, was sure to be a damp squib.)

In early October, Dr. Weaver and I spoke for about 45 minutes.

He described the virus's origins in Africa and its passage through Yap Island and French Polynesia, and he said very little work had been done on it thus far. The virus had arrived recently in Brazil and was worrisome because it was causing Guillain-Barré syndrome.

It might have spread even farther than Brazil, he said. No one knew. Only a few of the world's top labs could test for it, including probably only one in Brazil, the Oswaldo Cruz Foundation.

It was particularly hard to test for in Latin America, he said, because many people had previously had dengue or had been given yellow fever shots as children. Since they were related diseases, they produced antibodies that cross-reacted with Zika antibody tests.

If the disease ever came to the United States, he noted, it would at least be easier to test for, since dengue hadn't infected many Americans and only a tiny number had ever had yellow fever shots. (I was one, I reflected.)

Would it come? I asked.

It might, he said, but it would probably suffer the same fate as Florida dengue outbreaks, a cluster of a few cases that was crushed once it was detected.

"I don't think we'll see a major epidemic," he said. "We stay inside our air-conditioned homes."

He forwarded me a 2009 paper he had coauthored naming the viruses he thought were most likely to cross the ocean and hit the Americas. It was prescient: Zika was one of them. But it was among the also-rans. The two biggest threats he saw were Japanese encephalitis and African Rift Valley fever.

I thanked him, hung up, went through my notes to make sure I could read them in the future, scribbled his phone number and the date at the top, tore them off the legal pad, stapled them, and dropped them onto my head-scratcher pile. It's about six inches tall and consists of stuff that strikes me as interesting enough to write a story about. Someday.

And, for the next couple of months, that was that.

I went back to writing about curing infant jaundice with sunlight and the long-term repercussions for China of the reality that its males smoked one-third of all the cigarettes in the world. I was also trying to arrange a difficult reporting trip: first to Bangladesh to see the world's biggest diarrhea hospital, where crucial work had been done on a new cholera vaccine, and then to Vietnam to see whether it was true that communists are terrific at fighting tuberculosis, but would be able to declare victory only if they obtained more aid from the capitalists.

On Monday, December 28, 2015, recently back from that trip, I was in the office. It was the week after Christmas, and very quiet. No news was breaking, and half the editors were gone, anyway. My cholera and tuberculosis stories were coming along, but slowly.

I was poking through piles on my desk and wandering in infectious disease websites like ProMED and flutrackers.com. I had to dig up something to write for my

Global Health column, which was due in a few hours. Along with full-length stories, I do something every week for the Tuesday Science *Times* section. It's usually short, 300 words or so.

Cranking it out can be a pain, but it's also a useful outlet: there are many events or studies that are not big news, but still intriguing. For example, a new vaccine against leishmaniasis made out of sand-fly saliva. (Leishmaniasis causes festering wounds and was obscure until U.S. service members in Iraq started getting "the Baghdad boil.")

On Google News, I saw a small CNN story out of Brazil. It had "Zika" in the headline. Remembering the earlier conversation, I opened it—and read with growing horror.

Brazil had declared a state of emergency. Hospitals were seeing a wave of babies with microcephalic heads, more than 2,700 of them.

Zika was the suspected cause. The CNN piece mentioned the same facts Dr. Weaver had, but one line caught my eye: some of the country's top obstetricians, it said, were recommending that women not get pregnant. Another article I found said a health ministry official had advised the same thing.

That was mind-boggling. Outside of China and its one-child policy, I'd never heard of any government—or any sane doctor, for that matter—recommending that women just stop conceiving. The idea was a betrayal of the whole idea of nationhood.

I looked at the CDC's website. It had very little information: a paragraph stating that Zika virus was in Polynesia and South America, and that some cases had been reported in returning travelers. Nothing about microcephaly, nothing about Guillain-Barré. It did have one ominous line: "These imported cases may result in local spread of the virus in some areas of the United States."

I called the one person I knew in Brazil, an Italian doctor named Marco Collovati, who ran a diagnostics company.

A few years earlier, I had written a front-page story about a new rapid test for leprosy, which is a big problem in Brazil. His company had created it. I had interviewed several leprosy experts for it, but not him. Nonetheless, soon afterwards, I found a box on my desk with a ceramic figurine of a Brazilian gypsy. Attached was a note full of exclamation points: it was a thank-you gift. I emailed him to say thanks, but *New York Times* rules didn't let us accept gifts worth more than $25. When we spoke, he was effusive. "Dooooonald! You muuust take it! It is nothing! It is a souvenir! They sell them on the street! Your story has made me famous! I am a suuuuuper-hero in Brazil! You are a suuuuper-hero too! You are saving the world!"

Apparently, my story had been picked up by all the Brazilian media. Soon thereafter, he excitedly sent me a picture of himself on a dais with the very popular former president Luiz Inácio Lula da Silva, who was endorsing his test as an example of Brazilian ingenuity.

Marco was lots of fun, but that day he turned somber. I was incredulous. Was this story true? All these kids? From a mosquito disease? Yes, he said. His company was already working on a rapid test for Zika.

"It is a big, big mess, Donald. It is a tragedy. These babies do not recover. It is a very big El Niño this year, it is very hot. It is raining already, and it is only going to get hotter. The Olympics—they will be a disaster. Can you imagine people coming from the U.S., from France, into this?"

Yes, he confirmed, the health ministry had advised women not to have children. "What can they do? Abortion is illegal. So the only way to prevent this is to not get pregnant."

And was it definitely Zika?

"It's a big question," he said. "We don't know if it's only that, or maybe it's a combination. If a pregnant woman has had dengue a year before, gets an immunological reaction, and then gets Zika . . ."

I wrote a brief story with the basics, and after I filed it, I sent a long note to Tom Skinner, the chief spokesman for Dr. Thomas R. Frieden, the director of the CDC. It was headlined, apologetically, "Merry Christmas and let's make Tom Skinner crazy during Xmas season." If he had any days off, I was probably going to ruin them.

I had just written something brief about a new virus called Zika, I said. It was a huge problem in Brazil, suspected of causing a 20-fold increase in microcephaly, and women were being advised not to get pregnant. I was going to have to follow up with a major story, hopefully with help from the *Times*'s Brazil bureau.

When I do that, I said, I'll need to include the CDC's thoughts on how fearful Americans should be.

The agency's website indicated that some American travelers were getting the disease and bringing it home, and there would likely be outbreaks in the United States.

Americans were going to want answers to some questions, I wrote, including these:

1. Should Americans be concerned that Zika could spread in the United States and cause brain damage in children?
2. Should Americans avoid going to Brazil now?
3. What other countries should they avoid?
4. Should Americans cancel plans to go to the Olympics?
5. Should American athletes avoid the Games? Or should some subset of the team, like pregnant athletes or female athletes?

Could he please, I asked, put me on the phone with someone as soon as possible?

His email auto-reply said he was off until January 4. But he wrote back within 20 minutes, saying, "We should be able to make this work. Let me see who is around. What is your drop-dead deadline?"

Depends on my editors, I replied, but by tomorrow afternoon, please.

Three hours later, another CDC spokesperson wrote back saying Dr. Erin Staples, an epidemiologist in vector-borne diseases, would be available early the next afternoon.

In those three hours, my editors had heard from the foreign desk. Simon Romero, our Rio bureau chief, had interrupted his own vacation to work on a piece about the emergency. It would be offered for page one, so I could file paragraphs about the American situation into it.

Dr. Staples and I spoke the next afternoon. She described the risk to the United States—that it was carried by the same "yellow fever mosquito," *Aedes aegypti*, that carried dengue and chikungunya, so the agency expected the spread to be similar: Puerto Rico would be hit hard. There would probably be small clusters of cases in Florida, in Texas, and along the Gulf Coast—and possibly also in Hawaii. The CDC did not expect anywhere in the mainland to be hit as hard as Puerto Rico would be.

But, she warned, nothing was clear-cut: mosquito control budgets were set by states or counties, and they waxed and waned. They had gotten fatter when West Nile virus was a threat, but that was 15 years back, and West Nile was spread by different mosquitoes and had to be fought differently. The small dengue and chikungunya outbreaks since then hadn't moved the budget needle. And there was very little money for surveillance—which meant trapping and typing mosquitoes regularly—so nobody really knew the true range of the yellow fever mosquito. The maps were old. And there was another wrinkle: a related mosquito, the "Asian tiger mosquito," *Aedes albopictus*, could also transmit Zika. The tiger mosquito tolerated colder winters and survived much farther north.

On the other hand, she said, nobody was sure how "efficiently" the tiger mosquito transmitted Zika.

(Mosquitoes can be "inefficient transmitters" if viruses don't grow as well in them, or because they wander off to bite birds or deer or something else instead of humans. *Aedes aegypti* is highly efficient because it will live inside human houses, breeding even in shower drains if it can. And, unlike most mosquitoes, which bite and hang on until they are full or squashed, it is a "sip feeder"—it bites three or four times, very briefly, before going off to lay eggs. So it can spread diseases within a family rapidly.)

The rest of the conversation surprised me a bit. She was very noncommittal.

She didn't want to talk about Brazil and whether microcephaly was a threat. "Information is pretty limited, so it's hard to comment," she said.

For now, the agency was therefore sticking with its travel advice. Brazil, like all the rest of Latin America and the Caribbean, was at a level 1 alert, which was just "Avoid mosquito bites." That advice had been standard for a long time because of dengue and chikungunya. The CDC had no special advice for pregnant women. Nor for the Olympics. Just "If you go, don't get bitten."

It felt a bit blasé. If the CDC was worried, it wasn't showing.

Simon's story arrived, and it was powerful. It described pregnant women across Brazil in a panic. It quoted a health ministry official, Dr. Claudio Maierovitch, who advised that if women in the affected areas could wait to get pregnant, they should. The country normally had only about 150 microcephaly cases a year and was now investigating 2,782. Forty infants had died, and more were expected to.

It quoted Gleyse Kelly da Silva, a 27-year-old toll taker

from Recife. She had three kids, and then her youngest, Maria Giovanna, had been born with microcephaly in October. "I cried for a month when I learned how God is testing us," she said. "I had never heard of Zika or microcephaly. Now I just pray that my daughter can endure life with this misfortune."

Simon's story was up on the *Times*'s website on December 30 and on the front page the next day.

From that point on, the *Times* was driving the story forward. We wrote about it frequently, and the stories were often on the front page and prominently displayed on the website and mobile platform.

On January 4, 2016, the *Times* ran another article reporting that the United States was becoming more vulnerable to tropical diseases. Everyone blames global warming, but that was never the whole story. It was a combination of warmer weather moving mosquitoes north, of cheaper, more frequent jet travel helping more people reach new continents with viruses still fresh in their blood, and of the spread of urban slums like Brazil's favelas, where viruses that would have died out if victims had lived far apart benefited from the multiplier effect of people being crammed close together. It was compounded by bad mosquito surveillance and the use of pesticides that don't work anymore.

At the time, there had been about a dozen reported cases of Zika in the United States. All were in returning travelers, who had all recovered.

Then, on January 4, Puerto Rico reported its first case of locally acquired Zika—someone who had not been off the island recently had caught it. So now the disease was officially spreading on American soil.

Still nothing from the CDC.

(Actually, I learned much later that the CDC had noted the first Puerto Rico case on its website on December 31, 2015. But it kept Puerto Rico at its lowest travel alert level—"Take precautions against mosquito bites"—which was already in effect because of dengue. Its media office did not send out a news alert, and the case went unnoticed until an Associated Press reporter in San Juan heard about it from the health department on Monday, January 4, 2016, after the holiday weekend.)

By then, twelve other countries besides Brazil were reporting locally acquired cases: Colombia, El Salvador, French Guiana, Guatemala, Haiti, Honduras, Martinique, Mexico, Panama, Paraguay, Suriname, and Venezuela.

Many, like Puerto Rico, were tourist destinations.

I called Dr. Peter Hotez, the dean of the National School of Tropical Medicine at Baylor College of Medicine and a former president of the American Society of Tropical Medicine and Hygiene.

He was very nervous about Zika. "Nothing does this kind of damage except rubella," he said. Back in the 1960s, before Dr. Stanley A. Plotkin developed a vaccine, there had been an outbreak of rubella—German measles—that killed or crippled 20,000 American kids. Many had died in the womb, some had been born with microcephaly, many were born blind or deaf.

Should the CDC issue a travel alert? I asked.

Yes, absolutely, he said. "If my daughter was planning to get pregnant, I'd advise her not to go to the Caribbean."

"This is going to decimate the Caribbean tourism industry, Carnival cruises, et cetera," he went on. "Think about it—why do you go to the Caribbean? You do it to do things that will conceive a child, intentionally or unintentionally."

His medical school was in Houston, on the tropical Gulf Coast, and, even though it was January, he was already imagining the city in serious trouble come summer. He had spoken to the head of communicable diseases for the city health department, he said.

"I told him, 'You have to do something. You can't just wait until birth defects turn up in the labor and delivery suites.'"

"I didn't worry about Ebola coming to the U.S.," he said. "*This* one I'm quite worried about. If this spreads, it's a whole generation of neuro-devastated children."

I kept asking Tom Skinner why the CDC wasn't issuing a travel alert. Didn't the agency have a dengue-fighting operation in Puerto Rico? It was now on the front lines. Could I talk to the head of it?

That official had recently retired, he said. The other people in San Juan were mostly lab people, who couldn't comment. The vector-borne disease people were based in Fort Collins, Colorado. They saw this as like chikungunya: "There's nothing you can do to ever get rid of all the mosquitoes. We can just keep an eye out for hot spots and try to do something about them."

Well what are you doing now? I asked.

The lab people are leading classes for doctors so they can recognize the symptoms, he said.

I started calling cruise companies—Royal Caribbean, Carnival, and Princess—to ask whether people were canceling cruises or whether they had any plans to protect guests from Zika. The first responses I got were "Zee-what?" I had to explain. Then they said they wouldn't comment and I had to talk to the Cruise Lines International Association, which spoke for the industry.

A spokeswoman there sent me an anodyne statement saying the association was aware that the CDC had noted a possible link between Zika and birth defects. Travelers should check with their local health authorities before taking cruises, it said, and the cruise industry "routinely instructs passengers" about wearing insect repellent and long sleeves.

They did not want to discuss whether anyone was canceling reservations.

It was high season, and pregnant colleagues were asking me whether I thought they should go on planned Caribbean vacations. "I'm a reporter, not a doctor," I said. "You should get medical advice from doctors, right? OK? Good. Now that we've established that: No!"

On January 11, I asked our research department to help me find women nervous about travel. Within hours, I had emails and phone numbers for people who had posted questions on TripAdvisor, Twitter, or elsewhere asking other travelers for advice about mosquitoes in various places.

One woman I reached, Ashley, a 33-year-old lawyer, was very sweet. She was suffering a crisis of conscience: she had three young kids and was pregnant with a fourth, although it was so early that she wasn't ready to tell anyone. Her mother had plans to take the entire family—17 people—to the El Conquistador resort on Puerto Rico. Her mom had already switched the trip from the Dominican Republic because of chikungunya, and had paid $3,400 in change fees for that. And now there was Zika on Puerto Rico. Ashley was asking everyone she knew for advice because she couldn't get any from the CDC. Her mother had two days left before she would lose her deposit.

She had called her ob/gyn. "Normally, she's almost overcautious, so I thought she'd say no. But she laughed and said, 'Definitely! Go!' Then I talked to another OB, and she said, 'I wouldn't cancel.' I'm not sure what to do. I can't be the only pregnant woman going to Puerto Rico. That's why I posted the question on TripAdvisor. Someone wrote back to me. They said they had just been at the Conquistador, and they hadn't even seen a bug, and there were lots of pregnant women."

"I'm really struggling," she said. "Part of me feels—you have to live your life, so let's go. I could skip going and still get hit by a car or catch West Nile or someone could sneeze on me. And part of me feels: This is my baby! I have to protect it!"

We talked about what I knew about the virus and what experts like Peter Hotez had said to me.

The next day, she emailed me to thank me, and said she and her husband had decided to drop out and let the rest of the family go.

I asked her whether I could quote her—and could I please mention that she was pregnant, because otherwise the quotes wouldn't make any sense. She said she needed to talk with her husband. She called back later to say that, since it would help other women in her situation, it was fine.

(I later learned that El Conquistador, a spectacular Waldorf Astoria resort, was in Fajardo, on the island's eastern tip, and that, early in Puerto Rico's epidemic, Fajardo had more cases than any other place.)

On January 13, I got an off-the-record phone call that pushed the CDC into action. The caller suggested I look at the website of the Brazilian health ministry because it had posted something important. I looked it up and Google-translated it.

It was a report about autopsies on four babies; two were miscarriages, two were born full-term with microcephaly and had died within hours. All four mothers had had Zika symptoms during pregnancy. All four babies were positive for the Zika virus. In the microcephalic infants, it was found in their brains.

This was a *second* smoking gun. The Brazilians had found similar results on their own weeks earlier: Zika virus in the tissue or amniotic fluid of three malformed fetuses. But this report mentioned that the lab work was done for the Brazilians by the CDC.

But if the CDC did the work, it believed the results. This was important evidence. Why hadn't it posted it?

I called back the person who tipped me off. It was political protocol. The CDC had to wait for the Brazilians to announce. I said, They're standing on protocol in the middle of an epidemic? While American citizens are waiting for advice? That's ridiculous.

But the person who had called me had clearly done it at the request of someone near the top of the CDC. They were trying to force the Brazilians' hand.

I wrote to a CDC media person and said I needed to interview someone as soon as possible about the new report and whether it was going to make the CDC change its travel advice.

I also tried directly emailing the CDC's head of quarantine, Dr. Marty Cetron, whom I knew from hanging around at tropical medicine conferences, to see whether he would talk to me. He said Tom Skinner would get back to me.

Tom arranged a conference call with Dr. Lyle R. Petersen, the director of their division of vector-borne diseases.

Yes, the report from Brazil was accurate, Dr. Petersen said. It was "stronger evidence of the linkage."

OK, I said—so was the CDC going to issue a travel warning?

We're discussing whether or not to change the recommendations, he replied.

How can you *not* change them? I insisted. Pregnant women were making travel decisions right now, and you just told me you basically have proof that this deforms babies.

Tom stepped in. "We're *optimistic* that we're going to change them."

When? I said.

"We hope to have something Thursday or Friday."

It's been two weeks, I said. What's the holdup?

"We can't make these decisions in a vacuum," he said. "We're consulting with others outside the agency."

What others?

"There are a lot of possible implications," he said. "We have to implement this and get the word out."

He wouldn't say, but it was obvious. For tourism-dependent countries, CDC travel warnings are nightmares. Millions of tourist dollars would disappear. Those countries would scream if the CDC suggested anyone—pregnant or otherwise—avoid them. They would complain through the State Department and the White House. The time was probably being spent soothing ambassadors who were calling their presidents. The tourism industry would scream, too, but through its lobbyists.

So I wrote a story saying the CDC was debating whether to warn pregnant women and others against traveling. I put in Peter Hotez's hypothetical advice to his daughter; Ashley, who feared for her baby; and described the head-in-the-sand attitude of the cruise industry. I noted that an estimated 1.3 million Brazilians had been infected, that they were now investigating 3,500 cases of microcephaly, and that the Dominican Republic had joined Brazil in saying it would advise women to not get pregnant.

When I'd filed, I called Tom to let him know.

"You're writing something *today*?" he said, sounding dismayed.

Yes, I said. I'm not waiting for a handout press release while you guys make up your minds. I know it's happening.

OK, thanks for letting me know, he said, resignedly. "I'll give everyone a heads-up."

The story went up that evening.

But the next day—nothing. I bugged Tom: Press conference?

More likely tomorrow, he answered.

Seriously? I thought. Seriously?

The next day, Friday, January 15, was a circus. The CDC announced a noon press conference, then canceled it, then kept changing the hour. I hounded Tom. Just after noon, he said it would be posted on the website at one, and they would discuss it at two. "Go to CDC travel alert page and keep refreshing," he wrote. I kept doing that. Hours went by. My "refresh" finger got sore. Editors were getting antsy. The press conference finally started at seven that evening.

The agency issued "interim travel guidance." Pregnant women "should consider postponing" travel to Brazil, to any of the other twelve countries with Zika transmission, or to Puerto Rico.

Dr. Petersen described Zika as "a fairly serious problem."

The advice didn't suggest that anyone else change plans because of the Guillain-Barré threat. But it did have advice for women considering pregnancy: they should "consult a doctor"—which is sometimes a CDC euphemism for "get on the Pill" when they don't want to say it aloud because it might offend people opposed to contraception.

The alert was for entire countries, not just regions or cities. Dr. Petersen was asked about that and said the alerts applied to "most of" each country unless there was "specific evidence" that the virus was not being transmitted in a particular region.

That must have really upset some governments, I thought. A huge country like Mexico could have Zika in only one small region while much of the country, including mountainous destinations like Mexico City, remained totally mosquito-free. And pregnant tourists and pregnant businesswomen were still advised to avoid the entire country.

I wondered what was going to happen when Florida had its first locally transmitted case; by this logic, pregnant tourists from overseas should then avoid even Minnesota.

There were a lot of questions about the delay. Why had it taken so long? Dr. Cetron was on the line and said the alert affected a lot of countries and "we don't like to blindside partners." The CDC had to give advance notice so other countries didn't "hear about it for the first time in the media."

It had taken a while, but the U.S. government was finally taking the problem seriously.

The next morning, a new announcement drove home the consequences for Americans. The first American baby with microcephaly because of Zika was born. It was to a mother in Oahu, Hawaii, who had lived in Brazil the previous May, during her first trimester of pregnancy.

6

Fast and Furious

THE CDC's TRAVEL alert served as a global warning: This is dangerous. If you're pregnant, stay away.

People all over the world—and editors—were suddenly full of questions. What was Zika? What was microcephaly? What was Guillain-Barré? What would happen to the Olympics? What would happen during Carnival? What should people who weren't pregnant think? What were women who were pregnant doing?

And now that the CDC had spoken for America, what was the World Health Organization going to do for the rest of the globe?

Up to then, the WHO appeared to be lying low. It had left most of the responsibility for tracking Zika to the Pan American Health Organization, its branch in the Americas, headquartered in Washington. PAHO's website had much more to say about Zika than the WHO did.

Also, the WHO tries very hard to avoid issuing travel advisories. As a UN agency, it is a big members' club, and it answers to an annual convocation in Geneva of all the world's health ministers. They elect the director general. It is politically very difficult for that director general, as club president, to point a finger at any member and tell the world, "His country is contaminated. Don't go." It had been hard enough for the CDC to warn American citizens, even though that was its national duty.

So, two weeks later, when the WHO declared a global health emergency, it was almost anticlimactic. By that time, it was so obviously a crisis that one wanted to say, "Well . . . *yeah*!"

Moreover, the WHO declares emergencies awkwardly. So awkwardly that its bureaucrats rarely even use the word. They usually say "pikes," from PHEIC, a "public health emergency of international concern," although I have heard it pronounced as "picks" and even "fakes." The agency had previously declared just three in its history, because it had obtained the power only in 2007, when the Geneva assembly changed the house rules.

The first was in 2009, over the swine flu pandemic, the third was in 2014 over Ebola in West Africa. The second was an oddity; it was declared in early 2014, over polio. Polio was the opposite of an epidemic. It was on the brink of extinction. It had been held down to a few hundred cases a year globally for decades, but it had suddenly begun spreading again in Africa and the Middle East.

In the 2014 Ebola outbreak, Dr. Margaret Chan, the director general, had been accused by Médecins Sans Frontières and others of waiting too long to declare an emergency. She was clearly not planning to make that mistake again, but even in top gear, the WHO is ponderous.

It convened a committee of experts in Geneva on February 1 and invited various scientists fighting Zika to give evidence, all behind closed doors so that they could discuss data that was still unpublished.

Later that day, on the "pretty unanimous" advice of the committee, according to its chairman, Dr. David L. Heymann, Dr. Chan declared a PHEIC. She worded it very carefully: the emergency was not over the spread of Zika itself, it was over the *possibility* that Zika caused microcephaly.

In essence, it was a plea for scientists around the world to cooperate on answering that question instead of hoarding their data until academic journals felt like publishing it. And it was an official wake-up call to the health ministers of countries with their heads in the sand that they had a problem.

One of the things we Americans don't realize is how truly indifferent many other governments are to the fates of their people. We're used to our politicians reacting, even overreacting, to news, because they can remind voters at election time how quick they were. But many countries—whether nominally democratic, socialist, or communist—are run by elites focused on lining their own pockets or consolidating their power, and the possibility that their women—especially their poorest women—may have deformed babies doesn't really move them. If their wives and daughters get pregnant, they can always move to Paris or London. The WHO's leaders will never say that aloud about their club members, but they know it. Making those elites realize that a new problem is real, and not just something that Washington has cooked up for its own nefarious purposes, is part of the WHO's job.

On the ground, nothing really changed. The WHO does not have its own army of doctors, or an emergency fund. Instead, it requests help from national health agencies like the CDC and provides diplomatic cover when planeloads of them land in fragile but proud countries that might otherwise fear swarms of foreign doctors, many in military uniforms. That's how, for example, Cuban and American doctors ended up working side by side in Ebola treatment units in Africa.

The PHEIC did raise the disease's profile, although it was already pretty high. It also boosted scientific cooperation in a different way: many medical journals, from the famous *New England Journal of Medicine* to the relatively obscure *Cell Stem Cell*, began posting Zika-related studies online as fast as they received them. Most scientists cannot fight their instinct to hoard data, because publication makes their careers. But when the journals took the brakes off, rushing research onto the web, the knowledge gained became a de facto form of international scientific collaboration.

In fact, within a week, something extraordinary happened. All the major science journals signed a pledge called "Statement on Data Sharing in Public Health Emergencies." It bound them to make all articles they published about Zika available free online for the duration of the emergency, instead of charging their normal subscription rates, which could run into hundreds of dollars. Research funders—both public ones like the National Institutes of Health and private ones like the Gates Foundation—also signed it, promising to require anyone to whom they gave money to share data as fast as it could be written up.

One scientist set an example by going much further: he began doing his important experiments online, for anyone who wanted to follow them. David O'Connor, a pathologist at the medical school of the University of Wisconsin, had a colony of macaque monkeys. He infected several pregnant ones with the Zika virus and described their progress day by day. He posted their blood tests, amniocentesis results, and ultrasound pictures. One of the first things he revealed was that the high levels of virus in the monkeys' blood persisted for weeks—a very bad sign, since nonpregnant people usually clear the virus in about ten days.

The WHO did not declare Zika a pandemic instead of an epidemic. Its definitions of what constitutes a pandemic sometimes shift, but usually refer to a "novel" virus. Zika wasn't novel, since it was discovered in 1947 (although a new fight could eventually arise over that if genetic sequencing shows disease-altering changes, such as a mutation that made it more lethal). Declaring a pandemic would probably not have changed anything for the United States, but some countries have response mechanisms that are triggered by it.

Thankfully, no one asked the WHO to referee a fight over the name "Zika." As with "Sin Nombre virus," naming conventions had created a nightmare in the 2009 flu pandemic. That novel influenza virus had first been spotted in a pig-farming town named La Gloria in Mexico's state of Veracruz. (Which was unusual because, historically, new flus were first detected in Hong Kong, and it was assumed that they originated on the pig and poultry farms of southern China.)

Flus have been around forever, but they become "novel" easily because they have eight loosely connected genes that can be readily swapped between viral strains in a game of mix 'n' match. The new mix had two pig genes and arose in a piggery, so it was called a "swine flu" even though it was by then a human one, too. At that point the pork industry howled, saying the name was killing bacon sales. The name also sparked a crisis in Egypt, where the government slaughtered 300,000 pigs belonging to the Coptic Christian minority. It was religious prejudice, not public health; Egypt didn't have a single case of the flu then. It also had unforeseen consequences, since the Copts collected Cairo's garbage to feed their pigs.

Bending to pressure, the WHO stopped saying "swine flu." There was a long history of naming flus after their supposed places of origin—the Hong Kong flu, the Asian flu, the Russian flu, the Spanish flu. But a PAHO official strongly objected to "Mexican flu" or "Veracruz flu," or even "La Gloria flu," claiming that any place-name would demonize people from that place. Maddeningly, the WHO kept changing its institutional mind. First, it called the new flu just "H1N1," but that was confusing, because there is a seasonal H1N1. Then it was "A (H1N1) S-O.I.V." for type A (H1N1) swine-origin influenza virus. But headline writers and television anchors refused to touch that. Finally, it became "Pandemic (H1N1) 2009." But just calling it the 2009 swine flu has stuck.

The WHO emergency merely raised the appetite for stories. My colleague Catherine Saint Louis did one about microcephaly and its consequences for children, explaining that some were profoundly disabled, while some were called "microcephalic" but simply had small heads and were of normal or near-normal intelligence. The BBC did twinned profiles of two British boys close in age to each other. One could run and kick a soccer ball clumsily but could not speak words and had emotional struggles. The other had a smallish head and complained that it ached, as if his skull wasn't big enough for his brain. But he was articulate and apparently doing well in school.

French Polynesian scientists and colleagues in France released an important paper in the *Lancet*, a British medical journal. After reading the headlines about babies with microcephaly in Brazil, they had begun a hunt through the island's records of births and medical abortions after their Zika outbreak. They had found 19 cases of congenital abnormalities, including 8 of microcephaly, 7 of them in a tight cluster of pregnancies that had begun during four months when the epidemic was peaking.

My colleague Sabrina Tavernise interviewed Brazilian doctors about their first recollections, and did a story about mosquito control and how hard it was to kill *Aedes aegypti* because it bred and lived indoors with its victims, as cockroaches do, not off in the swamps, as some other species did.

From Brazil, Simon did one describing the tremendous surge the country had had in Guillain-Barré syndrome a year earlier, in early 2015, as soon as Zika had turned intense. It had been noticed by the health authorities, there had been worrying headlines in Brazil, and some scientists had noted French Polynesia's experience, but it had not alarmed the whole world. He found Patricia Brito, a 20-year-old bakery cashier who was in intensive care for 40 days and said it was "more terrifying than any horror movie," and Geraldo da Silva, a 43-year-old construction worker who said he had felt he was "drowning in a sea of mud."

Another rare Zika complication that was a footnote in the academic articles about Polynesia eventually made headlines because it caused the first American death. It was an unstoppable case of immune thrombocytopenic purpura, a tongue-twister of a name that means "purple skin caused by leaking capillaries caused by low platelets (thrombocytes) caused by an immune system problem."

The first American to die of Zika was not a baby but a Puerto Rican man in his 70s. He succumbed in February 2016, but the connection to Zika wasn't confirmed for two months. First, the health department and its CDC advisers had to find the antibodies in his blood—antibody testing takes much longer in tropical areas because dengue and yellow fever cross-react on the preliminary antibody tests, creating false positives. To distinguish them from each other, scientists must do "neutralization assays," a version of the same work that Zika's discoverers did in mice, but carried out in flat flasks of live cultured cells instead. It is faster than using whole mice, but still takes days or weeks. Then they had to dig up his medical records and interview his family and his doctors to be sure he hadn't had anything else.

The man was reasonably healthy for his age before developing Zika symptoms in January, quite early in the island's outbreak. He recovered, and everything looked fine. But a few days later, he demonstrated "bleeding manifestations," which the initial CDC report did not detail but, given the later diagnosis, presumably included blood leaking from his gums and nostrils as well as petechiae, tiny dot-like bruises all over his skin caused by leaky capillaries. If the bleeding doesn't stop, the dots grow until they merge, becoming purpura.

That would have alarmed him and his family. He went to a doctor, who hospitalized him. In less than 24 hours, he was dead.

Immune thrombocytopenic purpura (ITP) is related to Guillain-Barré, except that the antibodies triggered by the late immune reaction don't attack the nerve cells. They attack the platelets, which the blood needs in order to clot. Without them, one simply bleeds to death internally. If ITP had occured in Brazil, it had not appeared in headlines or in medical literature. By the time the death in Puerto Rico was confirmed, it had been declared the cause of death in three cases in Colombia.

But that was a very rare condition. Another frightening possibility surfaced when I was in conversation with Dr. W. Ian Lipkin, the famous virus hunter who runs the Columbia University Center for Infection and Immunity.

I asked him about the theory that microcephaly might be a consequence of an initial dengue infection that was followed by Zika.

He argued that microcephaly didn't need a viral one-two punch. Several years earlier, he said, his lab had given monkeys Bornavirus (a rare virus that attracted little attention until it killed three German squirrel breeders) and their babies had been microcephalic.

What people didn't realize, he said, was that microcephaly was just the tip of the iceberg. Regarding Brazilian kids whose mothers had Zika but who appeared healthy at birth, he said, "I wouldn't be surprised if we saw big upswings in ADHD, in autism, in epilepsy, and in schizophrenia."

That was a horrible thought, I replied. I'd assumed most damage would be apparent at birth. Didn't schizophrenia usually strike young adults?

Yes, he said, it would appear only later. But infections of the fetal brain could have a whole range of consequences. The gross ones were obvious; the subtler ones might emerge only in twenty years. "We're looking at a large group of individuals who may not be able to function in the world."

And there was no way to know which outcome was most

likely, he added. In 2010, his lab had infected pregnant mice with a synthetic RNA virus that replicated in fetal mouse brains. The results were wildly unpredictable.

"If you infected them halfway through gestation, the offspring got the mouse equivalent of depression—they were withdrawn; they sat in a corner of their cage and didn't interact at all," he said. "If you did it two-thirds of the way through, they were hyperactive, the equivalent of manic."

I consulted other experts on fetal brain development, and they agreed with Ian. Schizophrenia runs in families, so it is assumed to be linked to genes. It is also brought on or worsened by trauma such as sexual abuse, abandonment, or heavy drug use. But there was a growing body of evidence that infections in utero played a role. For example, adults born in late winter and early spring had schizophrenia more often than those born in summer and fall. The suspicion was that flu viruses their mothers caught shortly before their births did something harmful. The virus itself didn't have to reach the baby's brain, the theory went: even the "cytokine storm" that the mother's immune system generated in response could cross the placenta.

In 1988, Finnish researchers looked at schizophrenia rates in age cohorts and reported that it was quite high in the slice of the population born right after the 1957 Asian flu.

And a pioneer of schizophrenia research, Dr. E. Fuller Torrey, said he believed Rosemary Kennedy, President John F. Kennedy's older sister, had been a victim of the Spanish flu. Born at its height in 1918, she struggled to learn to read anything harder than *Winnie the Pooh* and in her late teens developed schizophrenia symptoms, including violent outbursts. At 23, she had one of the earliest prefontal lobotomies performed on her, and she was institutionalized until her death. Some historians had blamed it on oxygen deprivation at birth, but Dr. Torrey thought subtle damage from the flu epidemic was a more likely explanation.

Whether psychiatric problems are in the future of babies with Zika won't be known for years. As of now, though, doctors in Brazil and elsewhere are describing problems that confirm Ian's basic premise: fetal death and profound microcephaly are just the most extreme end of a whole spectrum of damage suffered by Zika babies. Some have normal-looking heads, but also both small empty spots and calcifications in the brain, produced when cells die and stop filling the spaces they should. The holes appear in different parts of the brain in different children, so some may have problems with decision-making while others may have trouble running or walking. Damage to the long nerves attaching the eyes and ears to the brain has also been observed.

Eventually, scientists from French Polynesia and France were to publish a follow-up study on the 19 pregnancies in which brain damage had been detected, usually on ultrasound. Eleven of them had been aborted. (In cases of severe malformation, French law permits termination of even a full-term pregnancy, as long as the scans have been sent to the Prenatal Multidisciplinary Diagnostic Center in East Paris and a committee there has approved.)

Eight babies were clearly microcephalic. Six were not,

but had brain lesions. Five appeared to be normal but, after birth, displayed problems that indicated brainstem damage.

Two have since died. Six that were still alive had been flown to France or Australia to be assessed and treated at top neurological hospitals. All six had "severe neurological impairment." They were born unable to suck, swallow, or clear their lungs, so they lived in intensive care units needing tube feeding and regular aspiration. They had other problems, as well, including epilepsy or irregular heartbeats, jaws and tongues that hadn't formed completely and thus partially blocked their airways.

Some of the mothers of the 19, the report said, had had Zika, but without symptoms. They had had no warning until they saw their ultrasound scans.

7

Sexual Transmission

BEFORE BRAZIL, EVEN before French Polynesia, the United States had a confirmed case of Zika. And it was one that would eventually make medical history, because it provided the first hint that Zika could be transmitted by sex.

It had happened in 2008. Brian D. Foy, a skinny, boyish-looking 36-year-old vector biologist and malaria researcher at Colorado State University, was on a trip to Senegal, in East Africa. He and one of his grad students, Kevin Kobylinski, 27, were gathering mosquitoes.

Photographs from their trip show them grinning and looking like ragged extras from the cast of *Ghostbusters*, wearing headlamps and backpack vacuum cleaners with long suction nozzles for aspirating mosquitoes out of the dark eaves of huts. They are standing on a wide dirt road in the countryside, and it is clearly hot. They are wearing cargo pants and T-shirts. "We were bitten a lot," Brian said later.

By coincidence, I had spoken to Brian in 2011, and later met him at a tropical medicine conference. His Senegal research was based on an idea I found fascinating: fighting malaria through xenointoxication, which sounds like intergalactic beer pong but is Greek for "poisoning the guest." The villagers from whose huts he and Kevin had been collecting mosquitoes were all taking a deworming drug, ivermectin. After it got rid of their worms, it lingered in their blood for weeks, and, like the worms, the mosquitoes that sucked their blood died of it, too. His theory was that, given often enough, deworming pills could wipe out malaria in an area. (Other experts considered the idea clever but impractical because the distribution of deworming pills to remote villages even twice a year was hard, and the pills also caused rare but dangerous side effects that had to be tested for.)

Shortly after Brian returned to Colorado, he and his wife, Joy L. Chilson Foy, a nurse and the mother of their four children, had sex. Six days after he got back, he fell sick, with a rash, fatigue, a headache, and swollen joints. He was sure it was something mosquito-borne he'd picked up in Senegal; he was even surer when Kevin fell ill with the same symptoms. But four days after that, Joy also fell ill. Her symptoms were similar, but worse; she had bloodshot eyes that hurt in bright light, a bad headache, chills, and an intense red rash.

Brian also developed other curious symptoms: pain between his legs and difficulty urinating, suggesting an inflamed prostate, and then a reddish-brown tinge in his semen that looked like blood. Joy "was not happy about any of it," he recalled.

But it was a puzzle: They were in northern Colorado. Joy had not left the United States in a year. There were no disease-transmitting *Aedes* mosquitoes in the Rockies, and even if there had been, it would have taken at least 15 days for a virus picked up by a mosquito to reproduce and make its way from the abdomen to the salivary glands.

Mosquito transmission seemed impossible. Perhaps, they thought, by a total coincidence, Joy had caught something communicable that was different but with similar symptoms. But there were no likely candidates, and whatever it was, their children didn't catch it.

Because Brian was a scientist, he had his own blood drawn, along with Kevin's and Joy's. Some went to the CDC for testing, and some he stored in his lab freezers.

The CDC tested for diseases known to circulate in West Africa. Brian and Kevin came up positive for antibodies to dengue. Joy did not.

But Brian and Kevin also might have had dengue from earlier trips. The CDC's answer was: "We think you two had dengue. We don't know what your wife had."

For more than a year, that's where the mystery stood, until Kevin went back to Senegal. While there, he had a beer with an entomologist from the University of Texas Medical Branch named Andrew Haddow. As luck would have it, he was the grandson of Alexander Haddow, one of the discoverers of Zika. After hearing the symptoms, he suggested testing the stored blood for 16 different local viruses, including Zika, and once he got home, he helped them do so through a UTMB laboratory that had the right reagents.

Brian and Kevin came up positive for dengue again and for yellow fever—which made sense because they had had shots. But all three, including Joy, came out positive for antibodies to Zika.

The only obvious explanation was sexual transmission. In 2011, Brian wrote a paper for the CDC journal *Emerging Infectious Diseases.*

"If sexual transmission could be verified in subsequent studies," he concluded, "this would have major implications toward the epidemiology of Zika virus and possibly other arthropod-borne viruses."

The paper described only the curious case of Patients 1, 2, and 3, but Martin Enserink, a reporter for *Science* magazine, realized that they were the first three coauthors and called to ask about their identity, and Brian confirmed it.

Once Zika became a household word, Brian gave a few TV interviews about the case. Joy did not join him on camera. The initial *EID* article had been illustrated with a photo of her bare rash-covered back. But she wanted her 15 minutes of fame as the poster girl for sexual transmission of arboviruses to end right there.

"This stuff gets sensationalized," Brian told me. "I'm fine with it, as long as it doesn't get too silly and crazy. I'm all about the science. But she's annoyed with the whole thing."

Since their case, only one other had raised even the possibility of sexual transmission.

It was in a relatively minor paper by a French Polynesian scientific team. Its members and colleagues in New Caledonia had been the first to discover that the virus was present at high levels in urine as well as blood (and that urine tests were more accurate than blood tests). The case described a 44-year-old Tahitian man who had recently recovered from a bout of fever that sounded like Zika but that he had not been tested for. He finally saw a doctor because, although the fever was gone, he had genital pain and blood in his ejaculate. Tests showed that he no longer had Zika virus in his blood, although he did have antibodies. He did have virus in his urine and even higher levels in his semen. It was not clear exactly what the source was; the pain and blood suggested an infection of the testicles or perhaps the prostate gland. The authors surmised that theoretically it was possible that he could pass the infection on through sex, though there was no suggestion that he had done so.

On January 25, 2016, I wrote an article about the two cases. The CDC had issued its travel advisory for pregnant women only 10 days before. The Zika pages on its website at the time had a brief note saying there had been two cases in which the possibility of sexual transmission had been raised, by the Foys and the Tahitian man.

I called some infectious disease experts to learn what they thought. Given how fast the disease was spreading, did this possibility warrant more alarm?

Dr. Weaver at UTMB said he thought the risk was high enough to worry about. If he were a man with Zika symptoms, he said, "I'd wait a couple of months before having unprotected sex. And if my wife was of child-bearing age, I'd want to use protection, certainly for a few weeks."

Dr. William Schaffner, the head of preventive medicine at Vanderbilt University Medical School said he thought just two cases were "not really enough to warrant a big public health recommendation from the CDC."

The world was still reeling from the CDC's travel warning on the basis of thousands of cases, and every week saw two or three more countries added to the list.

"But it's provocative," Dr. Schaffner added. "So someone else could recommend it. And it certainly should be studied."

Six days later, on February 2, 2016, there appeared another case of sexual transmission in Texas. Someone living in Dallas had gone to a country where Zika was circulating, had come back, had sex with a partner who had not left the country, and both had come down with Zika symptoms. There was only one explanation.

I called the Dallas County Health and Human Services Department, asking for more details. The spokeswoman was unhelpful, offering nothing. She finally said that the other country was Venezuela. That was it—nothing about how the county learned of the cases, how it had investigated: nothing.

Frustrated, I emailed Tom Skinner.

"Got anyone who can talk about this case of sexual transmission in Dallas? I guess someone's going to have to hand out condoms along with flyswatters."

He replied, "Call Dallas. We confirmed the test results but Dallas did the investigation."

I blew up. "Don't fob me off on Dallas," I wrote. "They don't issue national guidelines. My desk wants to make it page one. I'd like to talk to someone ASAP, please. This confirms what scientists have suspected—sexual transmission possible. CDC pages have always mentioned this possibility down at the bottom, but no guidance given out (i.e., condoms, abstinence . . .). Brits have counseled men to use condoms after traveling to Zika areas. Does CDC plan to issue similar advice?"

Three minutes later he wrote back, "We'll have a statement to you ASAP."

I should point out that I like Tom. I enjoy his good ol' boy joking, hearing about his family and his invitations to join him in what he does for fun—attend NASCAR races with a big radio scanner so he can listen in as the drivers debate tactics with their pit crews. He's very patient and good at his job. It's not his fault that he's stuck between his bosses and me.

An hour and 20 minutes later, the CDC issued an oddly garbled statement. It described the bare details of the case, and then reiterated that the best way to avoid Zika virus was still to prevent mosquito bites, and that travelers should avoid getting bitten on their return to prevent local outbreaks. It repeated its recent advice that women should postpone travel to Zika-hit areas.

But the statement did contain two sentences referring to sex. One read, "Sexual partners can protect themselves by using condoms to prevent spreading sexually transmitted infections." That sounded as if it been lifted from standard

CDC advice about syphilis and HIV. The other was quite odd for CDC-speak: "Pregnant women should also avoid exposure to semen from someone who has been exposed to Zika virus."

It concluded, "CDC will issue guidance in the coming days on prevention of sexual transmission of Zika virus, with a focus on the male sexual partners of women who are or who may be pregnant."

Three days later, on February 5, the agency did release more detailed guidelines. They suggested that men with pregnant partners wear condoms or abstain from sex for the duration of the pregnancy. For nonpregnant partners they did not specify for how long. (A couple of months later the CDC would refine those suggestions to eight weeks for men with no symptoms and six months for men with symptoms.)

In retrospect, part of the coyness surrounding the whole Texas episode may have stemmed from something not revealed at the time. When the case was described in the literature two months later, it turned out that both partners were male. The Dallas spokeswoman had studiously avoided using pronouns like "he" and "she." The CDC had taken pains to say that "in this instance there was no risk to a developing fetus," when it might have been phrased more simply as "the female partner was not pregnant." I don't know why health agencies were reluctant to admit that gay sex could transmit the virus—it's useful and pertinent public health information—but they went to absurd lengths to conceal it. As of May 31, 2016, the WHO reported 12 cases of probable sexual transmission in countries where mosquitoes were clearly not to blame, and three cases in which Zika virus was detected in men's semen. All the transmissions were from men to others. In one case, oral sex was suspected but not proven. One man still had it in his semen 62 days after recovering from his fever.

Sexual politics and CDC timing aside, the idea that a mosquito-borne virus could also be transmitted by sex was, for scientists, mind-boggling.

"This is a paradigm shift," Tyler M. Sharp, a CDC epidemiologist, said to me later. "I do arboviruses. I never thought I'd be working on an STD."

Viruses mutate constantly, which often shifts their virulence. Some influenza strains become lethal by producing proteins that jam a host's immune response, for example.

But one aspect usually remains fixed: how they are transmitted. Many viruses have spherical shells, but the shell surfaces are as different as those on tennis balls, BBs, and popcorn: they match receptors on the cells they infect, and a cell in the throat is very different from one in the gut or one in the vagina. Until recently, scientists believed that a skin virus never evolved into a sneeze virus and a bug-bite virus would never be transmitted by a subway pole.

Everyone understands mutation. A Great Dane can be mutated into a Chihuahua. But this was like a dog mutating the ability to fly.

Ebola had defied the stereotype, and that had been realized only two years before, though Ebola had been studied for more than thirty years. Ebola is transmitted by blood, vomit, feces, and contact with dead bodies. It is extremely lethal, but patients who recovered from infection had been considered safe. Then, toward the end of the West Africa outbreak, a woman came down with it, and the only logical explanation was that she had gotten it from sex. The outbreak was almost beaten, cases in the community were rare, and she had had no family contacts, funeral contacts, or anything else. Her one risk was that she had had sex with a former Ebola victim who had long before walked out of an Ebola treatment unit, apparently cured.

Stunned scientists finally figured out that he must still have had live virus in his testicles. They realized that Ebola could, in rare cases, break into the body's "immunologically privileged" zones—parts that are normally walled off from circulating blood and have their own sustaining fluids. They are not easily invaded by a virus, but once inside, the virus can replicate in peace, because it's equally hard for antibodies and white blood cells to get in and kill it.

The eyes are privileged, and the bizarre effect of Ebola breaking into one American victim's eye was that it turned his blue eye green for months. The testicles are also privileged, presumably to protect the sperm from damage that might be passed to future generations.

Zika, like Ebola, seems to be able to breach that defense. Sex is now considered the second-most-common mode of transmission of the epidemic.

But as of now, no scientific estimate of how often it occurs has been published. And many important questions remain unanswered: Can a man transmit it without ever having symptoms? Does blood have to be in the semen? Can a woman transmit it?

Scientists just don't know.

8

New York's First Case

THE FIRST KNOWN case of Zika in New York City was not connected to Brazil. It occurred in 2013, but little was made of it at the time.

And as with Brian Foy, it arrived in the blood of a young to middle-aged, highly educated white American male.

I make that point only because I get regular emails from readers saying things like, "I read your story today about Disease X. This just proves how illegal immigrants are putting us in danger. With our weak border policies, any one of them could be carrying it into the United States and threatening the health of Americans. Why don't you write about that?"

In truth, yes, immigrants do bring some diseases to this country. But so do Americans. When the emails aren't too rude (and ruder than me is a high bar), I answer them, pointing out that the 2009 swine flu spread through the eastern United States thanks to a group of students from a Catholic high school in Queens, New York, who went to Mexico on their spring break. The 1999 West Nile virus epidemic was almost undoubtedly sparked by a tourist returning from the Holy Land; the first cases were also in Queens (it's where JFK International Airport is), and the strain was identical to one circulating in Israel. And the last polio outbreak in America, in 1979, took place in Amish communities in Iowa, Missouri, Pennsylvania, and Wisconsin. One member of the sect had picked it up at a Mennonite convocation in Canada. The Amish have been Americans since 1760.

I heard about the case through an email from a public relations person representing the Jonas Nurse Leader Scholarship program. One of its scholars, a nurse-practioner named Dyan J. Summers, had written about it in an article for the *Journal of Travel Medicine* and at a conference for travel medicine specialists.

When I called, Dyan described how the patient had walked into her office at Traveler's Medical Service of New York, on Madison Avenue. He was a regular—a thin, fit 48-year-old who had just come back from a long trek with his wife that took them through Ecuador, Peru, Bolivia, Chile, Easter Island, and Hawaii, with a stopover in French Polynesia.

He pulled his shirt out of his blue jeans and peeled it off, revealing a pinkish rash he'd had for eleven days.

"I took one look and said 'dengue fever,'" Dyan recalled. "He said, 'I'm not so sure. I think it's Zika.'"

"I thought, 'What?' she said. "I'd *heard* of Zika. But nobody was *thinking* Zika. Nobody thought about Zika until this guy walked into the office."

"But you have to understand," she continued. "This is a very, very bright guy. He's very savvy, very well traveled. He knows about safe water, he takes his malaria pills and knows what altitudes are safe, he comes here for his pretravel vaccines. He was right on the money, that guy."

How had he known?

"In Polynesia, he read articles about Zika in the local paper."

She had snapped a picture of his back, and took blood samples then, and again twenty days later. The CDC testing protocols for Zika at the time said it could do a PCR test for the virus itself only if the blood was taken in the first 10 days of symptoms, and the patient was past that. Otherwise, it required two samples of "convalescent" blood taken at least two weeks apart so that it could do neutralization assays to compare antibody levels. The blood serum would be diluted again and again and then drops put on flat surfaces covered with cultured cells that had been infected with Zika. They would be checked every few days to see how many cells the antibodies had "saved" from death. If a barely diluted drop of blood from the first sample saved half the cells, and a very diluted drop from the second sample also saved half, then the antibody level in the blood must have increased.

In his case, because he often put himself in the path of mosquitoes, the traveler had antibodies to dengue, West

Nile, and Zika. But his Zika antibody levels had multiplied five times in the 20 days between the two samples, while the other two had remained stable. That was powerful evidence that what he was recovering from was Zika.

As they talked about the virus's ramifications, the conversation proved strangely prescient.

First, the traveler said he had found an article about a scientist in Colorado who had infected his wife—Martin Enserink's article about Brian Foy. Dyan called it up on her computer, read it, and advised him not to have unprotected sex with his wife. It was almost three years before the CDC issued the same advice.

"What's weirder," she said: "He knew there were cases of Guillain-Barré connected to it."

When he was there, in the last week of November 2013, Dr. Mons and her colleagues had probably seen fewer than a dozen cases. It would be a couple of weeks before local newspaper articles mentioned it, and months before doctors collected their data and sent it to medical journals.

I later met the couple, who live near Central Park. They asked that I use only his first name, Stephen, because they run an adventure travel agency under their married names. Even though he had completely recovered years ago, "people can get a bit freaky about exotic destinations," he said. "Googling me up with 'Zika' can make folks skittish."

He had a carefully clipped mustache and rimless glasses and was a software engineer—a meticulously organized guy with a taste for risky adventures, and clearly in good shape. He described an Oahu mountain, Koko Head Crater, he liked to climb. "It's a 700-foot Stairmaster, the local butt-kicker. Normally I just blast up that." Hawaii was their last stop before home, he said, and not being able to climb Koko Head because his back hurt so much let him know that he was really coming down with something.

They had been on a big loop trip lasting several weeks, most of it at high altitudes in the Andes where they hadn't worried about mosquitoes. But in Polynesia, the stop before Hawaii, they had stayed in a friend's large family compound, and had camped in the central courtyard. They had arrived wary of dengue.

"We had a tent—it was like a mosquito net on steroids, with a bottom and everything," his wife said. "So our nighttime exposure was zero."

"You spray up with DEET first thing in the morning," Stephen continued. "So the only time there was nothing on us was the few minutes when I was in the shower. I had two visible bites. And I killed a mosquito in the shower. It was bloody, so I knew it had got me."

The backache and fever kicked in a few days later, in Hawaii, eventually becoming so fierce that "Tylenol wouldn't knock it down even a percentage." A rash started on his shoulder, but at first he thought it was just an allergic reaction to his camera strap.

By the time they arrived in New York, his fever peaked at 103, which is high for Zika, although it hadn't been diagnosed yet. That, he thought, could have been anything, including a flu. He and his wife decided to wait another day before calling a doctor.

The next day, the fever had broken. "I knew my immune system had taken care of it," he said. "But then I looked in the mirror. From the waist up, it looked like I had measles. I said, 'Something's wrong.'"

"Yeah," his wife said. "It was the whole shebang."

They came into the clinic, thinking that, if it was dengue, they wanted to know quickly, because a second infection carried the risk of hemorrhagic fever.

But "it had been buzzing in the local paper that Zika was around," Stephen said. After they left, it turned out that two children in his friend's family compound had also gotten it. "So that's why I suspected it."

It was a good call. I later learned from reading PowerPoints by Dr. Mallet, the French Polynesia epidemiologist, that the week they were there was the outbreak's absolute apex. Doctors on the 76 inhabited islands were reporting new cases at the rate of 3,600 a week. The Guillain-Barré, Stephen said, he had heard about from a local journalist who was a friend, who had heard about it from a doctor or nurse and was still looking into it.

In Dyan's journal article about his case, Stephen, like Joy Chilson Foy, appears only as a rash-covered back. "I'm a curious person, so it's kind of cool to be my own scientific experiment," he said in February. "But now I'm Zika Man. So hey—I should get a costume!"

9

The Rumors

THE RUMORS STARTED just as the first alarm bells began to ring, well before the CDC issued its travel advisories or the WHO declared a public health emergency.

The pictures of the children in Brazil were so shocking that people seemed to have a hard time believing that an otherwise mild disease had done such damage. They reminded many of the aftermath of major disasters: radiation victims from Hiroshima, children deformed by thalidomide or Agent Orange or by mercury poisoning in Minamata, Japan.

The rumors were similar: the virus was not the real cause. The media was a bunch of gullible idiots. The real cause was X.

Some rumors I read about in other publications or by following links down the rabbit holes of the Internet. Some I learned about because readers wrote to me, saying more or less that I was the gullible idiot and should look into cause X.

According to the first rumor, the culprit was genetically modified mosquitoes released in Brazil to fight dengue.

Another put the blame on some form of chemical pesticide. The first version of that rumor that I heard claimed it was Roundup, the herbicidal weed killer. The second, which became much more tenacious, was that it was a larvicide put into standing water to kill mosquito larvae, including the drinking water barrels that millions of poor Brazilians had attached to pipes running off the tin roofs of their shacks.

A third set of rumors blamed it on vaccines. One version held that Brazil had imported a bad batch of rubella vaccine, so mothers were left unprotected, and rubella was known to cause microcephaly. Another version pointed to the new vaccine against pertussis—whooping cough—that Brazil had recently introduced.

Another rumor—which caused me a lot of difficulty because it was initially argued persuasively by a prominent Yale mosquito researcher working in Brazil—maintained that there was actually no surge in microcephaly cases at all. It was all just a big misunderstanding. Brazil, the argument went, had seriously undercounted its microcephaly cases for years. Now that a few hospitals had had clusters—and clusters are normal in statistics—the media panic had led the health ministry to alert doctors all over the country, who were now reporting every child with a small head. It was just a massive overcount.

For several weeks, I felt I was just putting out fires. Serious news developments were taking place, including the WHO's emergency declaration. But everything seemed to feed the rumors. For example, when Dr. Chan and Dr. Heymann announced the PHEIC, they emphasized that the emergency was not the rapid spread of the virus but the suspected microcephaly connection, and Dr. Chan's words were particularly cautious: "Although a causal link between Zika infection in pregnancy and microcephaly—and I must emphasize—has *not* been established, the circumstantial evidence is suggestive and extremely worrisome."

That caveat—that it was *not* established, after all those days of headlines emphasizing it—was jumped on by reporters and columnists and Twitter opinion leaders. Everyone wanted to prove he or she was too smart to believe the conventional wisdom. Every telephone press conference I listened to from Geneva or Atlanta had the same question over and over: "Do you *really* know that Zika causes microcephaly? What's the evidence? Some people say it's some sort of X—how do you answer them?"

The frustrating thing about telephone press conferences is that everyone is usually allowed only one question, and it was an embarrassment to the profession how stupid some of those "Some people say it's X" questions were. The best health reporters asked good ones, but everyone waiting in line got one brief turn, and then it was over. As each conference ended, I threw my headset off in frustration—which was easier on the office equipment than in the old days, when AT&T's stout receivers could put some serious scars in the paint of the *Times*'s gray Royal typewriters.

Each of the rumors had some kernel of truth that made it credible. And, as each one was debunked, another would take its place. Top health officials were tearing their hair out; they were trying to explain the science and warn people to protect themselves, and instead they were constantly being asked to respond to new proofs that the world was flat. Worse, in the countries themselves, each rumor made people in the path of the virus more dismissive. If the government says, "This mosquito disease is dangerous!" but a guy in the barbershop says, "Oh, I heard the Brazilians just panicked—you know Brazilians" or "I heard it's some American chemical in the drinking water—Monsanto, you know," then the other customers skip buying window screens so as not to look like chumps.

Something similar had happened during the 2014 Ebola epidemic: the initial outbreak was among the Kissi people in the interior where the borders of Guinea, Liberia, and Sierra Leone meet. But those three countries are run by elites in their capitals—in the cases of Sierra Leone and Liberia, by descendants of freed American and British slaves. They were dismissive of the "backwards Africans" in the interior, who distrusted them in return. So when word came down from the capitals that Ebola was a killer and people had to let their sick relatives be taken away by teams in space suits commanded by white foreigners and spraying bleach everywhere, and that they had to abandon deeply cherished and perfectly sensible customs like washing the blood and vomit off a body before a funeral or being able to lay a hand on a loved one to say goodbye, they rebelled.

The rumor spread that it was all a plot by the elites to soak the Europeans and Americans for money. People hid their sick and held funerals clandestinely. One medical team was even hacked to death with machetes. The epidemic spread partly because it took months to get average people to take it seriously.

This is sadly normal. Every new disease rides a wave of rumors. I had a long talk with Dr. Howard Markel, a medical historian at the University of Michigan. "Rumors are the lifeblood of every epidemic," he said.

He cited a whole series of examples. The Black Death in the Middle Ages was blamed on the Jews, who were accused of poisoning Christian wells. AIDS was initially blamed on the "gay lifestyle," including anal sex, intense disco dancing, and getting high on amyl nitrate poppers. His favorite was the rumor that spread during the 1892 cholera epidemic in New York City: it was the fish. Fishmongers' sales plummeted, so the fishmongers lobby leaned on the Board of Health, whose president held a public fish dinner to dispel the rumors.

But every rumor had some logic to it. Medieval cities had Jewish ghettoes, and plague sometimes struck there later. Not that the ghettos didn't have rats, but no one realized rats and fleas were the problem. Even rat diseases are spread by people—rats ride with loads of grain coming to market from outside the city, for example. Jewish and Christian markets might be separate; Jewish markets might not be as connected to the agrarian countryside as Christian ones were. Wells are dug in neighborhoods, so Jews and Christians often drew water from different wells. If one neighborhood was dying and another was not, the well-poisoning theory could seem plausible.

Fish are a logical target to blame for cholera. *Vibrio cholerae* is a water-borne bacterium. Sewage, not fish flesh, spreads these bacteria. But they do live in filter feeders like oysters, and New York's harbor in those early days was so famously full of huge oysters that they were a standard food of the poor. So, while fish were probably completely innocent, shellfish possibly were not, even though polluted drinking water was the real problem.

The "AIDS is a gay lifestyle disease" rumor was ridiculous, even more so when doctors realized within a year or two that the syndrome was the same as "slim disease," which was all over Central and East Africa. But it took another two years to find the virus that caused it. And rumors persist when prominent people endorse them. Peter H. Duesberg, a respected molecular biologist at the University of California at Berkeley, insisted for almost a decade that recreational drugs and the first HIV medicine, AZT, were the real causes of the symptoms and death. And more than a decade later, when the disease was widespread in South Africa, that country's president, Thabo Mbeki, read "AIDS denialist" websites and refused to let public hospitals offer antiretroviral triple therapy, saying it was a plot by Western pharmaceutical companies to sell pricey drugs to Africa. A 2008 study by Harvard researchers estimated that his policy had led to 365,000 deaths, including those of 35,000 babies.

There were so many Zika rumors, with so many facets, that my editors asked me to write one long piece wrapping them up and explaining why they weren't true.

The kernel of truth behind the mosquito one was that Oxitec, a British company founded by Oxford scientists, had bred a genetically modified male mosquito. It sought out and mated with female mosquitos but had a gene that shortened its own life and, more importantly, was passed on to all their offspring and caused 95 percent of them to die before reaching adulthood. (Oxitec was already modifying *Aedes aegypti* mosquitoes because they spread dengue, which had been raging through the Asian and African tropics for decades and in Brazil since 1981.) Oxitec had recently done field trials in Brazil, with the largest release taking place in Piracicaba. That created headlines because the words "genetically modified" make many people nervous, in Brazil as in Brooklyn. But Piracicaba is 1,700 miles from Recife, the microcephaly epicenter—about the distance from New York to Bismarck, North Dakota. Mosquitoes fly less than a mile in their lifetimes. Besides, the numbers the company bred and released were meant to cover a few neighborhoods. They were a drop in the ocean of billions, even trillions, of mosquitoes infesting South America. Also, male mosquitoes drink flower nectar, not blood. They don't bite people. Moreover, Oxitec had undergone earlier field trials—in the Cayman Islands, Malaysia, and Panama. There had been no microcephaly outbreaks.

The Roundup rumor I heard from a former newspaper colleague I hadn't seen in many years. She wrote a long passionate email to a neighbor of hers, who happened to be my former mother-in-law, who forwarded it to me.

"My conspiratorial reporter's brain," it began, "has been ruminating through the Zika virus panic about whether those birth defects might have another cause."

She had long followed environmentalists' efforts to get Roundup banned or labeled as a carcinogen, she said, and Monsanto was fighting back with "bullying" tactics, such as suing Hawaiian farmers who complained about its genetically modified seeds blowing into their fields.

Northeast Brazil, she said, had huge sugarcane and soybean fields where Roundup was used, "and I keep wondering whether the virus is being blamed for something that is actually being caused by the pesticide, which would really suit Monsanto."

Her hope, she ended, was that "the *Times* starts paying as much attention to the dangers of Roundup as it does to the dangers of a new virus."

Roundup also raised the GMO bogeyman because it worked differently from earlier herbicides. Like them, it killed broad-leaved plants, including most weeds, but did not kill plants genetically modified to resist it. Farmers using it had to buy Monsanto's "Roundup Ready" (meaning Roundup-resistant) seeds, creating a dependency that environmentalists found especially pernicious.

But the argument's weakness was that Roundup had been used all over America and much of the world since 1974 without triggering anything like what was happening in Brazil. Like all agricultural chemicals, it can be toxic at high enough doses. Farmworkers must take precautions, such as wearing gloves and not inhaling the spray mist.

But it had been sprayed on millions of acres for decades. Also, the Brazil victims were generally not workers on the giant farms of the northeast. They were students, ice-cream sellers, masons, bakery cashiers. Actually, many of them were residents of urban slums, where there was no space to grow anything and not an ounce of Roundup for miles around.

When I mentioned to Dyan Summers, the nurse-practitioner, that I was reporting this rumor, she burst out laughing. She has a twangy, sardonic way of wise-cracking that sounds like a young Dolly Parton. "My dad was a Roundup salesman," she said. "He used to come home reeking of the stuff. And I may be a little trailer park, but I am *definitely* not microcephalic."

The rumor about a larvicide came from a different source, though the Monsanto specter was raised again. It started when a group of Argentine doctors calling themselves the Physicians in the Crop-Sprayed Towns released a "report" blaming pyriproxyfen, a chemical that Brazil had been spraying into drinking water since 2014 to fight dengue. The report called Sumitomo, the Japanese chemical giant that made pyriproxyfen, among dozens of other products, a "Monsanto subsidiary," which it is not, although the two companies had collaborated on some research in the past.

Mark Ruffalo, an actor most famous for playing Dr. Bruce Banner and his alter ego the Hulk in *Avengers* movies, and also an environmental activist, was one of those who retweeted the Argentine report, helping it go viral.

Unlike some mosquitoes, female *Aedes* mosquitoes prefer to lay their eggs in clean water. (That may be how they got to the New World from Africa, by laying eggs in the drinking water stored on slave ships.)

It was true that some Brazilian states and cities had used pyriproxyfen for months or years to fight dengue, and that they sprayed it, or dropped pellets of it, into drinking water—into big holding tanks on hilltops and into personal rain barrels at the end of roof downspouts.

But pyriproxyfen is registered as safe for exactly that purpose. It's a chemical mimic of an insect hormone that signals larvae to stop growing. Insect hormones generally don't hurt humans, and vice versa. Creatures with internal skeletons, like us, diverged from exoskeleton creatures so long ago in the evolutionary past that each evolved different sets of signaling proteins. Bugs and humans both have legs, but they get very different chemical signals. That's why children in the American South in the 1950s could chase spray trucks, playing in the sweet-smelling clouds of DDT, while bugs flying through them instantly went into spasms and died twitching. DDT mimics the chemical that tells insect muscles to contract.

Moreover, local Brazilian officials said, some cities with microcephaly didn't use pyriproxyfen. They used alternatives like temefos, or they used nothing. And some cities that did use pyriproxyfen had no microcephaly.

Also, pyriproxyfen had been used in the United States since 2001. Under brand names like Nylar, Sentry, and Flee, it is still sold as a dog and cat flea treatment and as an anti-flea carpet spray. For 15 years, American babies have been crawling in it and putting their hands in their mouths.

The rumors blaming it on vaccines were routine. The antivaccine lobby is a constant presence and blames almost everything on them. Vaccine opposition is sometimes mistakenly assumed to have begun with the rumors about measles vaccine and the wave of autism that began in the 1980s. Actually, it goes back centuries. When Dr. Edward Jenner, sometimes considered the father of vaccines, published the results of his 1796 experiments, many respectable doctors were repulsed. He had stuck a lancet into a blister on the hand of a milkmaid with cowpox—a mild infection of the udder that humans can catch—took out some of the pus and pierced the arm of a young boy named James Phipps, deliberately giving him cowpox. After James recovered from cowpox, Jenner deliberately exposed him to smallpox, and he didn't get sick.

Jenner's discovery stands as a medical milestone, but in those days, most people still believed diseases were caused by bad air or an imbalance of humors. The "germ theory"—that disease was caused by creatures too small to be seen—was new, counterintuitive, and controversial, and many people rejected it. And the idea of deliberately sticking diseased pus into a child offended many average people, including many clergy, who railed against it as disgusting in itself and as defiance of God's will.

Blaming vaccines has become so routine that those rumors largely fell on deaf ears this time. There was no evidence that Brazil had bought bad rubella vaccine, because there had been no rubella outbreak nine months earlier. It was true that Brazil had relatively recently introduced a new diphtheria/tetanus/pertussis shot. But that was a change many countries, including the United States, had made years before. The new "acellular" component—made from broken-up pertussis bacteria instead of weakened whole-cell bacteria—was developed because the old vaccine had been blamed for occasionally causing seizures. It had nothing to do with microcephaly.

The rumor that it was all just an overcount was the tough one.

I'm not sure how it started—probably among scientists in Brazil, because some of them, even at the Cruz Foundation, definitely believed it.

I heard it first from Jeffrey R. Powell, a highly respected Yale professor of evolutionary biology who, among other pursuits, studies the genetics of *Aedes* mosquitoes. His lab did research in Brazil and had just shifted its focus to include Zika.

On January 28, he sent Simon Romero and me a draft op-ed piece he had written arguing that the microcephaly epidemic was a fiction.

The editors had declined it, he said, but he thought we might be interested in his thinking.

It was clearly a scientist's work, concise and packed with evidence. It noted that the virus had been in Africa and Asia for decades, apparently without ever causing a microcephaly surge. It noted that Zika antibody tests were unreliable in anyone who had had dengue or yellow fever, which many

Brazilians had. And its core tenet was that the Brazilian health ministry had, in the middle of its counting process, expanded its definition of a microcephalic head from one of 32 centimeters or less in diameter to one of 33 centimeters or less.

"This change in definition," he wrote, "increases by five-fold what is classified as microcephaly." His piece ended, "If we are lucky, in a year or so we may look back and conclude that the panic now occurring, most acutely in Brazil, was not warranted."

If he was right, Simon and I and the *New York Times* would look pretty stupid. We had been featuring the epidemic on the front page for a month, pushing it harder than other media outlets.

Even when they turned something down, the op-ed editors sometimes mentioned provocative ideas like that to newsroom editors. Also, submissions they rejected on occasion ended up in the *Wall Street Journal* or elsewhere. One way or the other, I was going to get quizzed about this.

A month earlier, when I'd started on the story, I'd read all the PAHO reports. I thought I remembered reading that Brazil's health ministry had changed its definition in mid-investigation. But what I remembered was that it had been changed in the *opposite* direction.

I dug through old reports until I found what I remembered, and sent Dr. Powell a note: "Unless I am misreading this WHO/PAHO page, the change in definition of microcephaly that Brazil made last December was in the opposite direction: Previously, newborns with heads less than 33 centimeters were considered microcephalic, now they must be below 32 cm. Normally, that would mean that far fewer children would be found to be microcephalic, no?"

Dr. Powell's first reply was that it contradicted what he had heard from Brazil, and he wanted to double-check. He later wrote back saying, graciously, "Well, seems I was wrong, and I thank you very much, Donald, for correcting me."

But it didn't end there. Wherever it had begun, the rumor was off and running. I was actually in California at the time, seeing my stepmother, who was declining from bone marrow disease, and then taking a break by driving down the coast. I was harder than usual for the desk to reach.

Then, on February 3, Brazil released the results of its first analysis of thousands of reported microcephaly cases. Another colleague in Brazil, Vinod Sreeharsha, covered it.

The results looked pretty damning:

Since the previous October, 4,783 cases of microcephaly had been reported.

The health ministry had thus far investigated 1,113 of them.

Of those, only 404 had been confirmed as microcephaly.

Of those 404, only 17 had tested positive for the Zika virus.

One might easily conclude that the skeptics were right: it was all a miscount.

I emailed Vinod. He was unhappy—his usual beat was business and political stories and, as a stringer, he lacked clout with the desk. Editors had read the report and worried that the earlier rumor was right. He had felt pressured to be cautious and emphasize the possibility that Brazil had over-reported cases. Other news outlets were being even more emphatic in saying the numbers implied it might all be a mistake.

Vinod had explained—correctly—how Brazil had tightened its microcephaly definition and had quoted both a Brazilian and an American expert calling that a medically sound decision. But those paragraphs were down near the end of the story, after a lot of copy stating an overcount was possible. Overall, the story served as a brief for the doubters.

The headline emphasized the skepticism: "Birth Defects in Brazil May Be Overreported Amid Zika Fears."

The next morning, Dr. Powell wrote me again, saying the article was exactly what he had suspected. Ambiguous microcephaly definitions and bad testing, he said, had "conflated to set off the whole hullabaloo."

And that was where things stood, for a while. The data was the data. The WHO had declared a global emergency just two days before. If the skeptics were right, it too would look foolish.

But I was sure the microcephaly was real, for one simple reason: in order to believe it was just a counting error, one had to assume that all the neonatal intensive care clinicians in at least four cities in Brazil's northeast were mistaken. In interviews that Simon and Sabrina had sent, and many others I'd read, they had all said the same thing: For years they had been seeing at most two or three microcephalic babies a year. Now they were caring for a dozen at a time in their wards. No neonatal specialist just fails to notice a deformed head. Also, the babies in the pictures didn't have heads just a centimeter or two below normal. They were truly tiny, and there were many pictures of them.

In retrospect, the Brazilian health ministry may have erred in reporting the results of its investigation so early. It was a public relations disaster. The ministry was opting for transparency, but the small percentage of confirmed cases and tiny number with detectable Zika virus made it look as if the agency had cried wolf.

A couple of weeks later, the ministry compounded its PR problems when it stopped reporting the number of unconfirmed cases. It made the change in an effort to squelch the rumors, but it looked like a cover-up.

Since then, the confirmed count has climbed to over 1,400.

It was no doubt true that Brazil had historically underreported microcephaly. With a population of 200 million people, it reported an average of 163 a year. In Europe and the United States, prevalence rates were at least two and maybe four times higher, depending on what definition of microcephaly you used.

Even so, that didn't come close to accounting for what had happened. Northeast Brazil, which is more sparsely populated than the south, normally reported 40 cases a year, a quarter of the national total. In just the six months from October 2015 to March 2016, the northeast states together reported 876 confirmed cases, nearly 90 percent of the national total.

And there was a sensible explanation for why the virus

was found in only 4 percent of the confirmed cases. Most of the mothers would have been infected in their first trimesters, six to nine months earlier. Antibodies usually wipe out live virus within two weeks. I was surprised there was live virus in *any* babies. It has since been noted, in blood tests on women and in the work on pregnant monkeys by Dr. O'Connor at the University of Wisconsin, that it sometimes does persist. How it does so is another medical mystery.

Those were the rumors as of early February 2016—and the answers. People would have to wait for more evidence, and then decide whether they found it persuasive.

10

The Proof

On February 1, 2016, when the WHO declared its emergency based on the *possibility* that Zika caused microcephaly, reporters asked WHO officials exactly what evidence was needed to be sure it did.

Initially, both Dr. Bruce Aylward, who was in charge of the response, and Dr. Heymann, the advisory committee chairman, gave the same answer: a large case-control study.

Scientists in Latin America, they said, were already recruiting pregnant women into one. They had signed up about 5,000, mostly in Colombia, some in Brazil, some elsewhere; they had to be pregnant and to have come up positive on tests for the virus. Those were the "cases." They were also signing up "controls": a roughly equal number of pregnant women who did not have Zika. They would try to match the two cohorts as closely as possible: same ages, same races, same neighborhoods, same income levels, same medical histories, especially regarding previous dengue or chikungunya infections. (Obviously, if a woman in the control group got Zika during the study, she would be shifted to the case group.)

They would monitor the two groups until their babies were born, and compare the results. If the Zika group had far more babies with microcephaly than the control group, they could definitively say Zika was the cause.

This was a "prospective cohort study," the gold standard in epidemiology. The women enrolled first were due to start giving birth in May and June, Dr. Aylward said, so final proof would have to wait until that data was ready.

In fact, the science moved forward much faster.

On March 31, without any fanfare, the WHO made a subtle but important change to one sentence on the face of its weekly Zika situation report. It read:

"Based on observational, cohort and case-control studies there is strong scientific consensus that Zika virus is a cause of GBS, microcephaly and other neurological disorders."

"Strong scientific consensus" marked a shift from previous reports, which said it was "highly likely" Zika was a cause.

Then on April 13, the CDC made it definitive. Its director, Dr. Frieden, scheduled an afternoon press conference with the leaders of his Zika team and declared unequivocally, "It is now clear: the CDC has concluded that Zika *does* cause microcephaly."

It was "an unprecedented association" in medicine, he

added. "Never before in history has there been a situation where a bite from a mosquito can result in a devastating malformation."

What led the agencies to change their minds about waiting for the big study?

A series of small studies.

The number of cases of confirmed microcephaly in Brazil had just kept growing. Before the CDC announcement, it had passed the 1,000 mark, with nearly 900 clustered in the northeast.

Microcephalic babies were by then being born not just in Brazil but in Colombia, in Panama, in Martinique, and in the Cape Verde Islands, not to mention the cluster discovered retrospectively in French Polynesia. Each cluster had followed a Zika outbreak about nine months earlier.

Separate teams of doctors—in Brazil, in the United States, even in Slovenia—had found Zika virus in the brain tissue or amniotic fluid of babies who had been born with microcephaly, had died in the womb with microcephaly, or had been aborted because microcephaly was detected on ultrasound. One particularly grim case was described in the *New England Journal of Medicine* on March 30. It involved a 33-year-old newly pregnant Finnish woman living in Washington, DC, who had taken a quick trip through Belize, Guatemala, and Mexico over Thanksgiving 2015. She had a routine sonogram on December 5, and her baby was fine. But she was having odd symptoms—a rash, eye pain, and a fever. Over New Year's, she was in Finland—presumably home for Christmas—and she must have read the news about Zika, which was just emerging then, and recognized her symptoms. She had a blood test and an ultrasound there. The ultrasound was normal. But her blood was positive for Zika virus. She went back to the United States and had the same two tests on January 5. Same results. Then, over the next three weeks, two things happened. Her blood remained positive for the virus, which was abnormal. And, horribly, her baby's brain began to dissolve. On her next MRI and ultrasound, at 19 and 20 weeks, the skull is the right size, but the surface of the brain has thinned out, the hollow spaces in the frontal lobes are larger than they should be, and the white matter that connects the two hemispheres is far smaller than it should be. At week 21, she decided to terminate the pregnancy. On autopsy, the brain was found to be teeming with viral particles. That was solid evidence.

There was also "biological plausibility," the CDC said. Biologists at Florida State University had tested the virus in several types of fetal cells that grow into the various parts of a baby. It barely infected some, such as the prekidney cells. But it homed in on the neural progenitor cells—the ones that ultimately turn into the brain—and destroyed them.

(Also, in results that were then still unpublished, when injected into immune-deficient mice, the virus did not kill adults but did kill fetal ones, and it was found in their brains.)

But the most convincing and most frightening piece of evidence was a miniature case-control study published by the *New England Journal of Medicine* on March 4. It was done by doctors at the Cruz Foundation in Rio working with a team from the David Geffen School of Medicine at UCLA. They described a group of 88 pregnant women whom they had started to enroll in September 2015, when the reports of microcephaly began coming out of the northeast. (Although the epidemic's initial epicenter was in that region, there was a simultaneous smaller outbreak in some parts of the Rio region.) The researchers were already in the middle of a dengue study, and they had noticed a few months earlier that they were getting many patients with fevers and rashes who tested negative for dengue. So they started testing them for Zika, and then began a substudy that looked just at the pregnant ones.

The study wasn't nearly over when they published the results. They had rushed it into print because what they were finding was so alarming that it needed to serve as a warning.

Rather than test all their subjects, they had chosen rashes as the recruitment factor. Whenever they saw a pregnant woman with a rash, they asked whether she would agree to participate. Eighty-eight had said yes. Of those 88, 72 tested positive for Zika. The other 16 became the "controls." Of the 72, 2 had miscarriages almost immediately. That didn't necessarily mean anything—miscarriages in early pregnancy are common. Of the 70 left, 42 agreed to have ultrasounds every few weeks. The other 28 refused. Some said the ultrasound clinic was too far away. But some "declined because of fear of abnormalities." That is, they preferred not to know whether their babies were deformed. They would find out at birth.

By the time the authors published their preliminary results, 12 of the 42 mothers having ultrasounds were showing evidence of "grave outcomes." Two babies had had normal ultrasounds, and then had suddenly died in the womb. Both of those mothers had been infected late in their pregnancies, not in the first trimester. The rest had ultrasounds revealing serious defects: some had microcephaly, some had white spots—brain calcifications—suggesting inflammation or cell death, some babies were much too small for their gestational age, some had almost no blood flow in their umbilical cords. By the time the study was published, 8 of the women had given birth, and the ultrasounds had proven accurate.

Twelve damaged babies out of 42 was a 29 percent "grave outcome" rate. The 16 Zika-free women acting as controls had zero problems. A difference of 29 percent versus 0 percent is more than "statistically significant." It's overwhelming. Among other things, those results forced experts to stop saying that the danger was all in the first trimester. Clearly, Zika could kill babies at any point.

"We were just blown away by that," Dr. Karin Nielsen-Saines, one of the authors, said. "We weren't expecting to find problems in all trimesters."

(A study done by the CDC that came out later, on May 25, in the *New England Journal of Medicine* looked at Brazil's northeastern Bahia state during the height of its epidemic and found that a Zika-infected woman's risk of having a microcephalic child was between one and 13 percent.)

The stack of evidence piled up by all these disparate studies, the CDC said, fulfilled "Shepard's criteria." That was a set of conditions published in 1994 by a professor of pediatrics at the University of Washington, Dr. Thomas H. Shepard, for determining whether a particular insult to a fetus caused a particular birth defect. (They were different from Koch's postulates, a better-known set of criteria published by the pioneering German microbiologist Robert Koch in 1890. But Koch's postulates are for concluding whether a particular germ causes a particular disease. Shepard's criteria relate to birth defects and incorporate nondisease causes like poisons or radiation.)

Dr. Bruce Aylward of the WHO was very pleased about the CDC's timing, even if it jumped the gun on waiting for results from the large study. It was both good science and "really responsible public health," he said.

It was good public health policy because many people in the Americas still doubted that Zika was the cause of microcephaly and were not taking precautions against it. As a result, babies would die. "If you're going to prevent disease, you've got to change behavior today," he said. "Not when it's too late."

By this point, the flurry of denial rumors had diminished, at least as far as I could tell. I was no longer getting emails about them. They had faded from social media, and the mainstream press was no longer repeating them. I wrote to Dr. Powell to ask whether his mind had changed. He replied that he was "becoming convinced there may be a causative connection between Zika infection and microcephaly." He did add that he was "less sure that mosquitoes are the sole culprit"—about which he was right, since sex clearly played a role, too.

Questions remained: Was the silence because of widespread acceptance? Or widespread indifference? And even if women accepted it, were they going to do anything about it? And what exactly *could* they do?

11

Delaying Pregnancy

WEEK AFTER WEEK, as the epidemic pressed outward from its Brazilian epicenter and more and more women fell within its orbit, one thing became clearer and clearer:

Nothing was stopping this virus. Not one country—not even one city or one island—was claiming that its babies were safe.

None of the vigorous mosquito-control efforts, none of the constant reminders to women to wear repellent and long sleeves, appeared to be winning. None of the calls to deploy genetically modified mosquitoes or bring back DDT were making any difference.

Much of that was wishful thinking anyway.

Oxitec, which bred GM mosquitos, was still in the field trial stage. The company would have needed thousands of insect hatcheries scattered all over the Americas to raise enough males to make a dent in the epidemic.

The spray trucks featured in so much television footage from South America were largely useless publicity ploys. Governments liked them because people found them reassuring. But against *Aedes aegypti* mosquitoes, relying heavily on street fogging was almost counterproductive: they bred near houses and slipped indoors as soon as they could, following the carbon dioxide vapor trail of human breath. As the trucks drove by, people closed their windows, thereby protecting the mosquitoes. TV footage of soldiers emptying standing water was also good publicity; but as soon as it rained, neighborhoods were back to square one.

And few of those calling for DDT to be revived seemed to realize that it would be a waste of time. Latin American governments had used it so intensely from 1947 to 1962 that they *almost* wiped out *Aedes aegypti* on the continent. In doing so, they *almost* eradicated yellow fever. But almost counts only in horseshoes and hand grenades. The genetic mutation that confers DDT resistance had emerged in Venezuela and had became fixed in the species before it spread outward from there. More than 50 years later, *Aedes aegypti* in the Americas still had the resistance gene. DDT was useless against it.

I kept asking mosquito experts to name one place I could go where mosquito eradication was demonstrably lowering infection rates. I usually heard long pauses, followed by "I can't think of one." It wasn't surprising. Dr. Frieden had said several times that there were examples from antidengue or antichikungunya campaigns in which mosquito populations, through herculean efforts, had been cut by 80 percent—with no effect on disease transmission. Ten to 20 percent of a mosquito population was enough to keep the virus circulating. Tests using traps inside homes showed that three mosquitos per household were enough, he said.

So how were women to avoid having microcephalic babies? To some Zika experts, the answer seemed screamingly obvious:

Women needed simply to not be pregnant. Not when the virus was peaking where they lived. Later—yes, fine, great, have children. But in the face of this unique epidemic, conception was uniquely dangerous.

And yet that insight was highly controversial for months. At the time of this writing, it still is.

For one thing, advising women to stop having children was unprecedented. Never in history had governments done so. (China's one-child policy was different: it was semipermanent and implemented for economic reasons.)

But the advice was also controversial because of the virulent reactions it provoked. Not just from the political right and the Catholic Church but from the left—from the very groups that were dedicated to defending women.

The controversy had begun as soon as the epidemic became known. The governments of six countries in the path of the virus separately made the suggestion.

In December, Dr. Claudio Maierovitch, who was in charge of Brazil's epidemic response, asked women in the northeast to postpone pregnancy if they could. Colombia and Ecuador followed suit in January, then Jamaica, El Salvador, and the Dominican Republic.

The length of their "suggested delays" kept growing. Alejandro Gaviria, Colombia's health minister, asked women to wait six to eight months. His Jamaican counterpart upped the ante: a year. Then El Salvador proposed two years—no babies until 2018.

Roman Catholic archbishops in each country objected. There was nothing wrong with "practicing self-discipline," as one put it, to prevent the birth of a deformed child. But if the health ministers were implying that women were supposed to use contraception, well, the church's opposition to that was well-known. It was using artificial means to frustrate the will of God. And abortion, the bishops said darkly, was of course out of the question. It was better to devote one's life to raising a handicapped child than to burn in hell for killing an innocent.

But that was predictable. The church's opposition barely made headlines. Churchmen were repeating what they had said for fifty years.

The more surprising, and much louder, backlash was from women's reproductive rights groups. They were angry because the advice came from men—not all the health ministers were men, but the first few to speak up were men from governments that had historically allied with the church.

The denunciations were furious.

They began with a January 22 Reuters story that was picked up around the world by everyone, ranging from the BBC to Fox News. It was actually a product of the Thomson Reuters Foundation, a charitable arm of Thomson Reuters that described itself as covering "humanitarian news, women's rights, trafficking, corruption and climate change."

The article was the first to take notice of the fact that multiple governments had offered the same advice. It mentioned Colombia, Ecuador, El Salvador, and Jamaica. Journalistically, that was a coup, and refreshing. At the time, most Zika articles were bogged down in all the rumors about the "real causes" of microcephaly.

After establishing the trend—governments asking women to wait—the author asked for reactions. But she quoted only representatives of women's groups. Not a single doctor appeared in the story.

Prominently featured was Monica Roa, chief of strategy for the Madrid-based women's rights group Women's Link Worldwide, who said, "It is incredibly naïve for a government to ask women to postpone getting pregnant in a context such as Colombia, where more than 50 percent of pregnancies are unplanned and across the region where sexual violence is prevalent."

In fact, contraception was free at Colombia's public clinics and abortion was legal in some cases. Roa acknowledged that, but said women had too little access.

That El Salvador's far more restrictive government was giving the same advice, she said, was "offensive to women and even more ridiculous in the context of strict abortion laws and high levels of sexual violence against girls and women."

Sara Garcia, a member of the Citizen's Coalition for the Decriminalization of Abortion in El Salvador, believed advice to delay had to include a public discussion of unwanted pregnancies. "There are pregnancies that aren't planned, are imposed on women and girls and are the product of sexual abuse."

American activists chimed in. "Once again, governments put the burden on women to protect themselves from any risk," said Paula Avila-Guillen of the Center for Reproductive Rights in New York City.

This point of view began to snowball.

NPR's *Morning Edition* did an interview with Roa, who called the advice "ineffective, naïve and unrealistic" because so many pregnancies were caused by rape and sexual violence.

Almost immediately, *Time* magazine did a piece titled "Why Latin American Women Can't Follow the Zika Advice to Avoid Pregnancy."

The article quoted Tara Damant, an Amnesty International activist who said governments were "putting women in an impossible position by asking them to put the sole responsibility for public health on their shoulders by not getting pregnant when over half don't have that choice."

It also quoted Avila-Guillen, who called the advice "naïve" and "irresponsible," noting that governments "were not issuing any recommendation for the men to use condoms, which is very unfair."

In the United States, outrage became the conventional wisdom.

Emma Saloranta, a founder of Girls' Globe, which shared women's motherhood experiences around the world, wrote a piece called "Zika Virus and the Hypocrisy of Telling Women to Delay Pregnancy."

The *Huffington Post* picked it up. The comments were almost uniformly favorable. Denying women both access and knowledge "sounds like a Republican wet dream."

It also struck a chord with right-wing outlets, which dislike governments' telling people what to do in their personal lives.

When I heard it, I began stewing. Women *needed* to avoid pregnancy somehow. Because clearly nothing else was going to save their babies.

First, it seemed inevitable even in January 2016 that all the known antimosquito efforts were doomed to fail. No country in the hemisphere except the United States had stopped dengue or chikungunya, and the United States had succeeded in part because it was rich and in part because even Florida had cold spells.

Second, given that, putting all the onus on women to avoid mosquito bites seemed absurd. No one can avoid them 24 hours a day for nine months.

Third, vaccine specialists had made it clear that there was absolutely no hope of a vaccine in less than two years at the earliest.

Fourth, the outbreaks appeared to be very short-lived. Yap's peaked and crashed in five months, French Polynesia's in seven. Neither country had reported a case since, and WHO categorized their outbreaks as "terminated." There were reports that northeast Brazil's and Colombia's were beginning to fade.

Fifth, if a woman wasn't pregnant, the disease was almost always mild. Getting it and recovering meant long-lasting protection. The disease itself was the perfect vaccine. And, if everyone around a woman was similarly "vaccinated," there was no virus for mosquitoes to pick up and infect her with.

What governments should really do, I thought, was ask women to wait if they could—and *encourage* everyone to get bitten. Yes, there would be Guillain-Barré cases, but better that than microcephaly cases. (The typical Guillain-Barré victim is a male of late middle age or older. Even WHO crisis guidelines, which favor minimizing the loss of what it calls "disability-adjusted life years," prioritize saving the lives of babies over saving those of old guys like me.)

Maybe, I thought, only half jokingly, people should donate blood to build up the supply of healthy plasma for Guillain-Barré victims, and then go get bitten.

Also, I was offended by the reproductive rights groups' rhetoric. I found it patronizing. In their scenarios, all women were victims and all men were monsters. Covering AIDS in Africa and elsewhere, I'd interviewed dozens of women about similar issues, and it wasn't that simple. Yes, there were teenage girls who couldn't avoid pregnancy. Yes, there were 40-year-old women who absolutely had to get pregnant. And yes, some men were monsters, and rape and sexual coercion were huge problems in various countries. But that wasn't the fate of *all* women. There were many married women with one or two children, who knew a doctor, who understood birth control. They had some power over their own bodies, and were able to say no, or negotiate a condom, or offer their partners another kind of sex. Presumably, their spouses or partners didn't want microcephalic babies either and would cooperate.

What had clearly gone wrong was that the health ministers had done a terrible job explaining *why* they were asking women to wait.

They had to realize they were not stopping the epidemic. Brazil and Colombia were already estimating millions of cases. They knew mosquito control was failing. Perhaps they didn't want to admit it. Perhaps they didn't want to insult those in charge of it—which was sometimes the army.

They also had to understand herd immunity. Most health ministers were doctors; they had studied the concept in medical school. But maybe they hadn't explained it in such a way that local reporters understood them. I'd read the stories from the countries concerned, and many of them were naïve. They quoted the ministers offering the advice, and then sometimes archbishops condemning it. But they hadn't asked, "Why?"

The debate had been hijacked; millions of poor women were being denied life-saving advice because it had become politically incorrect. I didn't see why women's groups had not taken the opposite tack. If they had embraced the advice, acknowledging that birth control and abortion would save women from misery, they could have used that as a wedge to try to get conservative governments to ignore fifty years of church pressure.

I had to see whether I was alone in this thinking. I started writing emails to virologists and public health experts, laying out my arguments above the final line: "Am I crazy? Or does this make sense?"

Almost universally, the answers came back: No, you're not crazy. Delaying pregnancy is good advice.

Dr. Marc Lecuit, a Zika expert at the Pasteur Institute in Paris who had studied Polynesia's outbreak, agreed. So did Dr. Weaver at UTMB. So did Dr. Albert I. Ko, a Yale School of Public Health infectious disease specialist who was helping set up a microcephaly study in Brazil. So did Dr. Ernesto T. A. Marques Jr., a Brazilian vaccine specialist who flew back and forth between Recife, his hometown, and the University of Pittsburgh's School of Public Health, where he taught.

Dr. William Schaffner of Vanderbilt went the farthest. Brazil had just announced that its army would join the fight. For one day, soldiers would go house to house looking for standing water and handing out pamphlets.

"They're mobilizing," he said. "Perhaps they should also be handing out condoms."

Everyone had caveats, of course.

"No government is going to say 'go out and get bitten,'" Dr. Schaffner said. "Because of the Guillain-Barré risk."

Asking people to hold off indefinitely would fail, Dr. Marques warned, because it would break up marriages.

He had another idea. When we spoke, in early 2016, everyone believed that the first trimester was the only dangerous period, so he proposed asking women to time their pregnancies so that their first trimesters did not fall in high mosquito season.

Also, the more I dug, the more the data confirmed that the reproductive rights groups had exaggerated women's helplessness. The Guttmacher Institute ranked countries according to how much access married women had to modern birth control.

Some Zika-hit countries ranked very low: in Guatemala, Bolivia, and Haiti, less than 35 percent could get it.

But others did well: in Colombia, it was 73 percent; in Brazil and the Dominican Republic, 70 percent; in El Salvador and Paraguay, 61 percent; in Ecuador, 58 percent.

Admittedly, those figures were for married women. Add teenagers, and the rates would drop. But that's true in every country, including the United States. Teenagers usually start having sex before consulting a doctor and are lucky if they have even a free condom from the basket outside the school nurse's office handy.

And Latin American and Caribbean women clearly had some power to choose. Fertility rates—lifetime births per woman—had been dropping across the region for two decades, just as they had two generations earlier in heavily Catholic European countries such as Italy, Spain, Portugal, and Ireland. The church was still effective at fighting abortion. But it had long ago lost its grip on birth control.

I also learned that there was a historical precedent for using epidemics to win reproductive rights. Long before *Roe v. Wade*, the 1964 rubella epidemic had played a role in the American abortion debate. That epidemic damaged 20,000 babies. By 1968, four states had passed laws permitting termination of a pregnancy if a serious birth defect was suspected.

On February 5, 2016, I wrote an article headlined "Growing Support Among Experts for Zika Advice to Delay Pregnancy."

It was on the front page of our science section and had a fair number of readers.

But it got no traction at all. Nobody openly disagreed with it. Nor was there any discussion. It just died.

The CDC and WHO continued issuing the same advice: avoid mosquito bites. Use DEET, wear long sleeves. Hold tight while we work on a vaccine.

Off the record, however, people in the CDC and its overseer, the Department of Health and Human Services, told me the issue was splitting their agencies. It had become, they said, a debate between two camps: the infectious disease specialists, who felt that asking women to delay was the only way to save babies, versus the reproductive health specialists, who said the government should not tell women what to do with their bodies.

But no one in the infectious disease camp would be quoted disagreeing with CDC policy.

Dr. Frieden later acknowledged that debate, saying my article and a set of questions I had sent him had been passed around and "triggered a long conversation."

In late February 2016, I went to Puerto Rico, since it was a piece of America right on the front lines, where the CDC was spearheading its efforts.

Zika was just beginning to take hold on the island; there were only about 100 confirmed cases. It was not even close to the point where everyone knew someone who had had it. Nonetheless, it was expected to overrun the island eventually. *Aedes aegypti* was everywhere. Serosurveys showed that 90 percent of all Puerto Ricans had had dengue, and 25 percent had had chikungunya, even though the latter had been there only eight months.

That didn't surprise me. I attended Zika classes given at WIC clinics—Women, Infants, and Children sites that give out what used to be called food stamps, teach breastfeeding, and offer other services. Since 92 percent of all pregnant women in Puerto Rico visit them, the government deemed them the perfect place to distribute information, insect repellent, mosquito nets, and condoms, and the CDC had created a 20-minute PowerPoint presentation for them.

"Ladies, this year's fragrance is DEET," the instructor at one class I visited said as she held up a green can of repellent—a blatant knockoff of OFF! Deep Woods."We all should smell like this."

But Puerto Rico had been putting out scare messages about mosquito-borne diseases for years, and fatigue had set in. When I spoke to women from the class afterward, I got very different reactions.

The first was 21 and newly pregnant. She was scared for her baby and so wore ankle-length dresses. When I pointed at her open sandals, she said she wore repellent under them.

"I take baths in the stuff," she joked. "I put it on in the morning and in the afternoon, and again when I sleep. And my mother is crazy with the bug spray."

The second, by contrast, was 30, in her third trimester, and wearing a tiny pink top and short shorts.

"You're not exactly mosquito-proof," I said.

"I know," she said, smiling and putting a hand over her cleavage. "I should cover up more. But it's hot."

She burned citronella candles at home, she said. Her father had come over to clear her rain gutters, and she had a neighbor with a "flea machine" who had fumigated her house as a favor. She was making an effort, but it wasn't going to protect her.

A third student said she never bothered with repellent because her two-year-old said it burned. Why not just wear it yourself? my translator asked her. She shrugged. "I didn't think of that." She lived on the 16th floor of a nearby housing project, and "mosquitoes don't go that high," she said.

Outside, I asked the instructor whether she didn't feel she ought to set an example. She was wearing a short white medical coat and red high-heeled sandals.

"Puerto Rican women are *not* going to stop looking good," my translator—a good-looking Puerto Rican woman—interjected.

Was she wearing DEET? I asked the instructor.

"Oh, not today," she said. "It smells. I usually wear pants."

Then she dropped her voice a little, embarrassed. "I should," she said. "I'm pregnant. We just found out."

That moment—meeting a well-educated, caring woman in the path of the virus who was so familiar with the threat that she was teaching classes in it, and who was in her first trimester of pregnancy and yet too busy or too . . . something to follow life-saving advice—convinced me that all efforts to protect pregnant women were just pointless. If even *she* couldn't be perfect for nine months, nobody could.

"What are you going to do if you get Zika?" I asked.

"I *won't* get Zika," she said firmly.

"OK. But if you do?"

"If that happens . . . I will have to face my baby's reality."

"What does that mean?"

"The greatest percentage of women who get Zika do not get microcephaly."

"OK, as far as we know from Brazil and Polynesia, you're absolutely right. But if you did?"

She said, very calmly, "I would face my baby's condition."

I knew abortion was legal in Puerto Rico. It was available in major hospitals too; women didn't have to go to a clinic with shouting protesters outside.

"You wouldn't . . . ? Consider . . . ?"

She shook her head.

That happened several times in Puerto Rico: women would not only avoid discussing abortion; they often wouldn't even enunciate the word.

I also interviewed several of the country's top obstetrician-gynecologists. Some said they were privately advising patients not to get pregnant. It was too risky.

A TV-star doctor, Dr. Jose Alvarez Romagosa, a fertility specialist who headlined a show called *Latin Doctors*, told me that he'd dissuaded three patients that day from conceiving. His partner, Dr. Hiram Malaret, said he had stopped inducing ovulation because he was worried about the babies—and the malpractice suits.

Dr. Manuel Navas, a hospital director in Fajardo, which had some of the earliest Zika infections, said he was discouraging all his patients. That was the advice he would give his daughter, he said.

When I asked what advice the Puerto Rican government was giving, I got contradictory answers. Some said it had kept silent on the issue. I'd noticed that there had been no discussion of it during the WIC classes I'd sat through.

Some, on the other hand, said the island's health secretary, Dr. Ana Ríus, had given a radio interview saying women should wait. But she had run into a buzz saw. The archbishop of San Juan had attacked her, and a popular radio host had accused her of being alarmist. She had turned shy and dropped the subject, they said.

I met her right before I left. She spoke softly and did appear to be shy. But she was adamant: her position was still that women should wait until more was known about the disease. She had asked the WIC clinics to hand out condoms for that reason, she added.

I told her I had been to clinic classes, and they weren't conveying her message. They had said the condoms were to stop sexual transmission. "I didn't know that," she said. "Thank you. I'll talk to the lady in charge of them."

Did she get a hard time from the archbishop and a radio host?

"Yes, I was criticized. But I haven't changed my message. I am a very Catholic person, but for me, public health goes above the norms that the church makes."

"Besides," she said, smiling. "I'm backed by the pope."

(Pope Francis had recently implied that condoms might be acceptable under the "lesser of two evils" doctrine, saying Pope Paul VI had permitted nuns in the Belgian Congo to use birth control because so many were being raped during the liberation struggle.)

Why hadn't her views been disseminated more? I asked. Why no big TV and radio campaign, billboards, newspaper ads?

There was no money, she said. Puerto Rico was broke. All she did was hold a weekly news conference to update the case figures and answer questions.

On March 8, 2016, the WHO issued an advisory echoing the CDC's. It suggested that pregnant women avoid traveling to areas where Zika was spreading.

During the telephone press conference afterward, I asked, not very politely, "If you're telling pregnant women not to *visit* countries with Zika because it's too dangerous, why aren't you telling women who *live* in those countries not to get pregnant? It seems inconsistent."

Dr. Heymann, the advisory board chairman, answered, "We don't give national recommendations." Dr. Chan, the director general, added, "We respect the law of the land."

I had known David Heymann for years because he was in charge of polio eradication at the WHO when I started covering it. He had a long, noble history as a disease fighter, helping eradicate smallpox and running some of the first WHO teams tackling Ebola outbreaks. He had been close to the top of the agency but had left to chair England's Health Protection Agency and teach.

I emailed him later asking for a clearer answer. Both pregnancy and nonpregnancy were legal everywhere, so "respecting the law of the land" made no sense. Besides, the advisory had advocated making birth control widely available, which did flout some country's laws.

We ended up in an email conversation that lasted several days. He said women had to make the final decision. I said of course they did, but they needed clear medical advice. He said he and Dr. Chan had hesitated because birth control must be voluntary, and can be abused. China enforced it, and India awarded poor men radios for getting vasectomies.

He was right that in Africa, Latin America, and parts of Asia, birth control–related aid from Geneva or Washington can be controversial, whether it's Norplant to prevent conception or condoms to prevent AIDS. It is often seen as white people trying to stop brown people from reproducing. To avoid that charge, the WHO avoids the term "birth control." They call it "birth spacing," and emphasize the health benefits to the mother of "spacing" children.

I said we weren't talking about reducing childbirths but about delaying them, perhaps only briefly, to prevent lifelong misery. I made all the arguments I'd made before, sent him my February 5 article, and told him what I'd seen in Puerto Rico.

Finally, he said, "OK, you've convinced me." But he would have a hard time winning over Dr. Chan and others, he added.

Then he said, "Would you like to co-author an article in the N.E.J.M. making the argument?"

That was a shock. I'm a jackal of the press with no medical degree. I've learned on the job by interviewing a lot of smart people and reading their work. I'd never been invited to do anything like write for the *New England Journal of Medicine.*

I said I was honored, but I had to check with our standards editor. The editor said no, a *Times* reporter couldn't ethically cover a debate and write a paper advocating one side of it. It was frustrating, but he was right.

On March 25, 2016, the CDC modified its guidelines for pregnant women. It did so in light of growing evidence that the virus persisted in semen for weeks.

To women *visiting* Zika transmission areas, it gave highly specific advice: Any of them wanting to get pregnant should wait eight weeks after their return before trying. If their partner had symptoms, they should wait six months. If she was already pregnant, they should avoid unprotected sex for the entire pregnancy.

But for women *living in* Zika-infested areas, the guidance was painfully wishy-washy. Timing pregnancy was a "very complex, deeply personal decision," the guidelines stated. Women should consult their doctors.

The doctors in Puerto Rico had told me how frustrating they found this. Patients were terrified. The CDC said, "Talk to your doctor." But it gave the doctors little guidance. They felt it passed the buck. Hundreds of thousands of Puerto Rican women of child-bearing age were left groping in the dark. Florida, Texas, and other areas would probably soon be next.

A CDC "Zika summit" in Atlanta was coming up. State and local health officials were invited, as were reporters. The former so they could share strategies and hear the CDC's latest thinking. The latter because the agency and the White House wanted to reinforce the message that Congress needed to vote the $1.9 billion the president had requested for fighting Zika.

As part of my reporting on the summit, I spoke to government doctors involved in the debate. None would publicly disagree with the CDC line because the Obama administration, as many White House correspondents have pointed out, very much dislikes internal dissent aired.

One doctor, who saw patients part-time at a clinic where almost everyone was on Medicaid, was very frustrated. You can't just give *hints* about life-altering decisions, she said. "Patients need to see the advice in black and white."

Another doctor was just livid. "The CDC guidelines are bullshit! Bullshit!" he shouted. In discussions of the February 5 article, he said, "some people write you off because they don't think Puerto Rico is like Yap. It's not like an island in the South Pacific. It may be more like Brazil—the epidemic will smolder, not disappear. Same with Florida, even more so. But they fail to see that, even if you're wrong about that, what you say has validity. One-third of all pregnancies *are* planned. Those babies *could* be saved."

He had hoped the American College of Obstetricians and Gynecologists would come out in favor of counseling patients to wait, he said, but their new Zika guidelines were also "milquetoast."

Dr. Laura Riley, director of labor and delivery at Massachusetts General Hospital and the guidelines' chief author, said the college's members were also split. Privately, some were suggesting delay to patients, and she had tried to let the guidelines include that, "but in ways that aren't proscriptive."

"Telling women what to do in the midst of an epidemic is difficult," she said. "For some people, 22 or 24, if there's an ability to wait, and you give them the tools, it makes sense. Our guidelines talk about making contraception available and talking to the patient about them. But if you're 41, it's not practical."

The CDC, to which she was a consultant, "is stuck with the political situation," she said. "Even if they said 'Delay,' by the time it worked its way up through HHS and government, those lines are going to come out."

If politics were holding the CDC back, I asked, couldn't her college, which had both private and government doctors as members, push them in that direction?

"Medical societies don't want to come out with recommendations different from the CDC."

Later, Dr. Jeffrey S. Duchin, chair of the public health committee of the Infectious Diseases Society of America, used much the same language. The IDSA had not discussed the idea, he said, and even if his committee took it up and decided delay was sound advice, it was controversial, and he doubted the society would want to "get out ahead of the CDC."

At the Atlanta summit, I sat next to Tom Skinner. I asked who the philosophical leader of the reproductive health camp was, and he pointed to the woman in a blue military Public Health Service uniform then at the podium, Dr. Denise J. Jamieson, an ob/gyn and team leader in the CDC's women's health and fertility branch.

Dr. Frieden confirmed that he was "guided by Denise's perspective as an ob/gyn" and added, "Here's one thing I've learned at CDC: if you're in disagreement with Denise Jamieson, you're probably wrong."

I asked for an interview, and got permission for one. Every such request at the CDC has to be cleared. The interview quickly turned tense.

I went through my thinking about the unstoppable nature of the epidemic and the possibility that avoiding pregnancy, at least during the transmission peaks, could be the only way to prevent microcephaly. Why, I asked, would the CDC not advise women to wait?

"Some countries recommended that during the 2009 flu pandemic," she said. "It was very poorly received."

I covered that flu pandemic, I said. I never heard that. What countries?

She had been working in South Africa then and couldn't recall right now, she said.

Nonetheless, I said, delay would prevent microcephaly.

"Yes, it would," she said. "But you'd also prevent *wanted* pregnancies, like those in women getting older."

Since when do older women want deformed babies? I said.

"Most women in Zika-endemic areas will have healthy babies. The majority will."

"Yes, but some *won't*. I don't see why you say it's 'not a

government doctor's job to tell women what to do with their bodies,'"

"That's right. It's not."

"Well, why do women go to gynecologists, except for advice about what to do with their bodies? Don't you tell women who are marrying HIV-positive men to use condoms? The CDC keeps saying pregnancy is a 'very complex, deeply personal decision that only a woman can make for herself,' right? But so is a double mastectomy. That's deeply personal advice about what to do with your body. Of course, the woman can only make it for herself. But if she has cancer, and you're her oncologist, and you don't tell her, 'Ma'am, you need to have this operation or you'll die,' aren't you guilty of malpractice?"

She stayed firm.

"I think the government getting involved in highly personal decisions about when to have a baby is not likely to be very effective," she said.

"Suppose you were in your job in 1964, and you knew that huge rubella outbreak was starting. There was no vaccine. You knew the consequences. Babies would suffer. What would your advice have been then?"

"I'd say, 'This is an extraordinarily risky time to get pregnant.'"

"But you won't give the same advice now?"

"This is different. There was no vaccine then. Highly motivated women can avoid mosquito bites."

"For nine months, 24 hours a day? Is that realistic?" I described what I'd heard from pregnant women in Puerto Rico.

"Well," she said, "that gets to the limit of our ability to make recommendations."

"To women who *visit* Puerto Rico and the Virgin Islands you issue very time-specific guidelines—wait eight weeks, wait six months, etc. Those are mostly white women, tourists. But to the women who *live* there, who are brown and black, you only say, 'It's a very personal decision.' Their kids may end up deformed. What do you say to the view that that's racist? That, since it potentially kills babies, it borders on genocidal?"

"We'll give the same advice in Florida if and when there is ongoing transmission."

"I spoke to ob/gyns in Puerto Rico. They say they want leadership on this, and there isn't any."

"This *is* leadership. What we're saying is, 'It's an individualized decision. It's not the same message for everybody.'"

No matter how aggressive and rude the question, Dr. Jamieson answered it. She didn't dodge. And she stuck to her principles and her argument. Journalistically, I had nothing to complain about.

On April 14, 2016, I wrote another article, entitled "Health Officials Split Over Advice on Pregnancy in Zika Areas."

By this time, outside experts were even more aghast at the CDC's reluctance.

"It's a no-brainer," said Peter Hotez, dean of the Baylor tropical medicine school. "They should say, 'Don't get pregnant—watch TV for six months and you won't have a badly hurt baby.'"

Houston had just had a flood and was swamped in water. He published an op-ed piece in the *Times* saying how dangerous the summer was likely to be. CBS News interviewed him, and he took the camera crew around a poor neighborhood, showing them old tires full of water that were mosquito havens.

He told me that he had specifically said during the interview that women in Houston should consider not getting pregnant. CBS decided not to use the statement. He had said the same thing when *PBS NewsHour* interviewed him on April 18, and PBS did use it.

As of this writing, the CDC and the WHO are unmoved. Dr. Aylward acknowledged that the WHO was having its own internal debate—"theoretically, many have thought it may work"—but it was not going to issue official advice. There were too many unknowns, he said, including how long to wait. In a country of 200 million like Brazil, the epidemic would not necessarily fade away as it had on islands, and the risk period was now the whole pregnancy, not just one trimester.

PAHO was leaning more toward nodding approvingly as individual member countries suggested it. Dr. Marcos Espinal, who was running that agency's response, said he "did not have a problem with waiting three to six months," as Colombia had suggested. "But I do have a problem with two years," as El Salvador had, he said. "You're changing a whole generation."

It could be argued that thousands of children with birth defects could change a whole generation to an even greater extent.

12

The Future

WHAT NOW?

No one knows for sure. The epidemic is still emerging. Predicting how viruses will behave is a fool's errand. Predicting people is worse.

But some things are clear:

Zika is very much on the move. Transmission is increasing in Central America and the Caribbean, and will keep doing so at least until the fall. The disease is now headed for its first summer in the United States, where no one has immunity.

That doesn't mean an explosion is inevitable. The CDC expects "limited clusters" anywhere in Florida or in parts of Alabama, Louisiana, Mississippi, and Texas close to the Gulf of Mexico. Hawaii is also considered vulnerable. That has been the national experience thus far with dengue and chikungunya, which are carried by the same mosquitoes. There have been pockets of cases in Key West and Martin County in Florida, in Brownsville, Texas, and on several Hawaiian islands.

But the outbreaks stayed small because most Americans, even in poor neighborhoods, have screened windows and air-conditioning. From a mosquito's standpoint, we live in nearly impregnable castles. American children don't usually get dozens of bites each night as poor children in Brazilian slums do, so the mosquitoes can't spin up a viral whirlwind by transferring it frantically from neighbor to neighbor.

The other reason earlier outbreaks never spread was that the authorities sprang into action fast. In 2009, it took only three dengue cases in Key West for the Florida Keys Mosquito Control District to roll out its helicopters and truck sprayers, to send teams down the island's streets with pesticide foggers and backpack tanks that shot larvicide into pools of water, to disperse other teams that went house to house asking residents to check their birdbaths and gutters, chlorinate their pools, and drop larvae-killing pellets into everything that collected rainwater. It was an impressive effort; although the outbreak ultimately lasted two years, it was held to ninety known cases. Even more impressive: the first case was in faraway Rochester, New York, in a woman who kept going back to her doctor saying, "I don't feel right," even after the doctor had diagnosed and treated her problem as a urinary tract infection. Eventually, on her third visit, the doctor consulted an infectious disease specialist, who suggested a dengue test because she had visited

Key West—even though dengue had not been seen in Florida since 1934.

However, Zika is different. Many dengue victims and 80 percent of all chikungunya victims see doctors quickly because they have high fever, headaches, and joint pain. Outbreaks are spotted early.

But 80 percent of all Zika cases are silent, and many symptomatic ones are mild. People often ignore early signs for days and call a doctor only when they look in a mirror and see bright red bloodshot eyes and a chest covered with a rash.

As a result, outbreaks may spread widely before anyone calls the mosquito teams. The more that happens, the thinner the teams get stretched.

Zika's spread may end up more closely resembling West Nile's. That virus also has silent cases. It's unlike Zika in that it's carried by *Culex* mosquitoes, which live all over the country. Also, it must simultaneously circulate in birds, whose hotter blood amplifies the virus enough for new mosquitoes to pick it up and infect humans. Nonetheless, despite helicopter spray flights and plenty of scary public service announcements when it arrived, it proved unstoppable. It entered the United States in New York City in 1999 and made its way slowly but steadily west for six years. All that held it up was winter. It moved a few states west each summer, then had to wait for the birds and mosquitoes to come back. It didn't really infest the Pacific Northwest until 2005.

West Nile is now endemic in the United States. It circulates every summer. About 2,000 cases are diagnosed each year, and about 100 persons die of it; the typical victim is a man over 65. Occasionally, there are sudden outbreaks, like one in 2012 that killed 69 people in Dallas–Fort Worth, pushing that year's death toll to a record 286.

Even if something like that happens with Zika, there will probably never be a huge surge of microcephaly in the continental United States. If West Nile caused brain damage in 1 in 1,000 cases, then 2 babies would be harmed each year. (One-in-1,000 odds is a very crude estimate from Brazil, where it was estimated that 1.3 million infections occurred in 2015, and the country has had more than 1,400 confirmed cases of microcephaly. But cases are still being confirmed and infection numbers in Brazil are still growing, so the ratio could change.)

But there is no guarantee that Zika will follow that pattern. There are too many uncertainties. How far *Aedes aegypti* will range this summer will depend on how hot and wet the weather gets. *Aedes albopictus* mosquitoes will range farther, since they tolerate lower temperatures, but whether they will aggressively spread Zika is still unknown.

If the virus is ever going to hit hard, this summer will be its best opportunity, since virtually no one is immune. If it persists and becomes endemic like West Nile, each summer's outbreaks will be limited by the growing portion of the population that is immune.

On the other hand, if it does that, it will never completely go away. Even if many women choose to hold off getting pregnant this summer, that can't last forever, so they will eventually be at risk. Their best hope will be a vaccine.

Dr. Stanley A. Plotkin, inventor of a rubella vaccine, predicted in January that making a Zika one would be relatively easy. Vaccines against other flaviviruses, including yellow fever and Japanese encephalitis, already exist. So it should be possible to take the "spines" of those vaccines, he said, and just attach Zika antigens, the proteins that provoke the immune system to make the right antibodies.

A majority of Americans—55 percent—polled by the University of Pennsylvania's Annenberg Public Policy Center in May said they would be likely to get a vaccine if there was one.

At least 18 private and public research labs are working on vaccines, from the Butantan Institute in Brazil to Bharat Biotech in India to a partnership of South Korea's Gene-One Life Science with Philadelphia's Inovio Pharmaceuticals. Some are producing "killed" versions and some "live, attenuated" ones. In the former, the virus is grown in cells, killed with heat or a chemical like formalin, and purified. In the latter, the human virus is weakened by one of several methods. It can be passaged through monkey cells or chick embryos or something else nonhuman; it can be forced to grow in nonhuman conditions such as low temperatures, or a piece of its genome can be snipped out or silenced. Once injected into a human, it reproduces for a while, but slowly, until the immune system dispatches it. Live vaccines provoke the strongest immune responses, but are also the riskiest, because they do grow and, in very rare cases, mutate into something threatening. They are not usually used in pregnant women or immune-compromised patients.

The National Institutes of Health has three vaccines in the works, said Dr. Anthony S. Fauci, who oversees them. One is a killed vaccine, one is a live attenuated, and one—the farthest along—is a new technology called a DNA vaccine. In that one, bits of the Zika virus's genes are spliced into a plasmid, a small ring of DNA that can break into cells the way a virus does. Once inside, it generates "virus-like particles," which have enough pieces of the Zika virus for the immune system to react to, but aren't whole virus and so can't cause disease. The live attenuated one is a "chimera," so named after the mythical beast with three heads: lion, goat, and snake. The scientists took an already-weakened dengue vaccine virus, snipped out the genes coding for the viral shell, and inserted those genes from the Zika virus into their place.

As of this writing, the DNA vaccine is supposed to move into the human-testing phase by September, and the two others are to follow within about six months after that.

That doesn't mean any vaccine will be ready soon. The most optimistic scenario Dr. Fauci predicts is two years, if everything goes right. Most experts expect three to five. Pessimists say twenty to never.

Testing proceeds in stages. Phase I is safety testing to make sure no completely unexpected side effect shows up. In this case, it will probably be in about 80 healthy adult volunteers from the area around NIH headquarters in Bethesda, Maryland. They'll be watched for three months.

By January, if that goes well, testing will move to a country with active Zika transmission and perhaps 2,000 or so volunteers will be recruited, initially just healthy adults. How long it takes to see whether a significant difference emerges in the number of infections between those who got the vaccine and those who got placebos will depend, Dr. Fauci said, on how intense the transmission is. In a raging epidemic, it could appear in months. In one that has faded, it could take years.

Meanwhile, safety testing will begin on children, the elderly, and perhaps even pregnant women. Those involve tougher ethical decisions. Normally, pregnant women are the absolutely *last* group any vaccine or drug maker agrees to experiment on. If it harms babies—and some experimental vaccines, notably an HIV one, have actually *increased* infection rates—the moral guilt is a bottomless pit. Not to mention the lawsuits. But pregnant women are the whole reason for making a Zika vaccine. Most likely a killed vaccine will be tested in them, Dr. Fauci said, but live ones will not.

If the epidemic has completely died out everywhere—which seems extremely unlikely—then the last options are "viral challenges" and "monkey models."

In a "challenge," some healthy volunteers who have been vaccinated and then tested to be sure they have antibodies will be deliberately injected with Zika to see whether it protects them. It can probably be done ethically because it's usually a mild disease. (I say "probably" because an ethics board will have to weigh the Guillain-Barré risk.) Ebola vaccines aren't challenge-tested because Ebola kills. Malaria vaccines, however, are—because there is a cure for malaria. If the vaccine flops, you can still rescue your volunteers.

Because of modern PCR testing, scientists can now use monkeys, whereas scientists in 1947 could not. Even if all monkeys remain visibly healthy throughout testing, if the virus builds to high levels in the blood of unvaccinated ones after a challenge but does not in vaccinated ones, it works.

However, some experts, like Michael T. Osterholm, who runs the University of Minnesota's Center for Infectious Disease Research and Policy, harbor doubts that there will ever be a vaccine—because of Guillain-Barré. Some vaccines, in rare cases, also trigger it. If it is a side effect in the range of 1 in 4,000, as French Polynesia's outbreak suggested, many thousands of volunteers must be recruited to make sure the vaccine clearly does it less often than the disease. Even a random Guillain-Barré case or two from any cause will scramble the testing statistics.

There are other ways to fight a virus, of course.

The most practical solution would be a treatment, or even a cure. But to hunt for a cure, you must know exactly how a virus does its damage, and science is only beginning to figure that out for Zika by trying to re-create every stage of the infection process in mice. The placenta is a series of semipermeable barriers between the mother's blood and the baby's, and it's not clear how the virus breaches those barriers, but a new study found 1,000 times as much virus in fetal mouse placenta as in the maternal blood. It is also not known exactly how the virus attacks the growing brain, but more than one recent study has suggested that it targets radial glial cells and may break into them through a surface receptor known as AXL. Radial glial cells appear very early in the brain-formation process and resemble snakes that have swallowed prey; they are long and stretched very thin but with a bump in the middle where their nuclei are. Their long, thin tentacles connect layers of the brain, with "feet" at one end and hairlike cilia at the other. They are believed to be the scaffolding that guides into place other cells, like astrocytes and neurons, that will ultimately form the brain, so any injury to them is devastating. AXL receptors are common on radial glial cells, and they are also thought to be the means through which Zika enters skin cells—which suggests that there may actually be a common thread to two very different Zika symptoms: skin rash and fetal brain damage. But the science is very new and nothing is yet certain.

But even if the mechanism is figured out, there is no guarantee that a cure can be found. There are many antibiotics that kill bacteria, and many chemotherapeutics that fight cancer. Unfortunately, relatively few drugs kill viruses.

A tumor is just a regular human cell that has gone haywire and is growing far faster than it should. Many drugs kill human cells; the trick is finding drugs that kill fast-growing ones while sparing healthy ones. Bacteria are practically animals: they ingest molecules, they make proteins, they move, they even catch viral infections. They have many processes that can be interrupted.

Viruses, on the other hand, are just shells containing bits of RNA or DNA that hijack cell machinery. Hepatitis C is one of the few viral infections that drugs can actually cure; for those patients, antivirals like Harvoni are miracle drugs.

Some drugs merely slow viral replication down, as Tamiflu does influenza or Atripla does HIV.

There is no proven cure for flaviviruses like Zika. In virology journals, many papers describe flavivirus "inhibitors" that work in cell cultures. Most are obscure chemicals that turned up when whole libraries of chemical compounds were screened. One paper published in 2012 was intriguing; it said ivermectin was highly potent against yellow fever and somewhat effective against Japanese encephalitis. William C. Campbell and Satoshi Omura won half of the 2015 Nobel Prize in Physiology or Medicine for inventing ivermectin, and it is the cornerstone of campaigns against worm diseases like river blindness in Africa. It is well-known to American pet owners; it's the active ingredient in Heartgard for dogs. (It was also the drug Brian Foy was investigating in Senegal.)

Other papers say various hepatitis C drugs work against Zika in the lab, as well as chloroquine, an antimalaria drug discovered in 1934, and amodiaquine, a newer antimalarial. The last two are both known to be safe for pregnant women. Many drugs may have potential, but none have been approved yet.

In theory, another route would be monoclonal antibodies. During the 2014 Ebola epidemic, much fuss was made over ZMapp, a cocktail of three cloned antibodies that had completely protected monkeys in tests. By the time the outbreak ended, however, it had been given to only seven people, two of whom died. Its maker said supplies had run out.

Antibodies are expensive to manufacture in bulk. ZMapp

was produced from a chimera of mouse and human genes, then grown in tobacco plants, extracted, and purified in tiny amounts. Scientists spent many years developing it. Work on Zika would have to start from scratch.

Immunoglobulin—antibodies derived the old-fashioned way, from human blood—has been used to treat Zika-related Guillain-Barré.

It has been used occasionally in pregnant women with Rh-negative blood and histories of multiple miscarriages, to protect the fetus. But it carries other risks and is cumbersome to get and administer, so no one has yet suggested a role for it in pregnant women.

CDC guidelines for doctors treating pregnant women with confirmed Zika are heartbreakingly simple. Doctors should offer "supportive care," like headache and fever relief, and ultrasounds and MRIs every three to four weeks. That—and encouragement to hope for the best—is all they have.

When microcephaly began making news in January 2016, doctors said it usually could not be detected before the third trimester, about the 28th week of pregnancy, which put women under tremendous time pressure while they struggled with an agonizing decision over whether to terminate. That time has been reduced—there are cases where fetal brain abnormalities were picked up as early as week 19. But it is not predictable, and no one knows how soon after an infection damage will appear.

Because no one in this hemisphere is immune, PAHO has predicted that the virus will reach every country except Canada and mainland Chile. In the South Pacific, it is still spreading to new island groups, such as American Samoa and Fiji, causing intense outbreaks. It also reached the Cape Verde Islands, which lie in the Atlantic between Brazil and Africa and are Portuguese-speaking; the virus is the same as the one in Brazil. It has even gone as far as the Maldives in the Indian Ocean.

The initial worries that it would circle the globe, however, seem to be getting more remote each day. Europe and most of northern Asia don't have *Aedes aegypti*. The worry was that the Cambodian-Polynesian-American strain of Zika would be so different from earlier Asian and African strains that they would not be protective. But surveillance has been conducted since January and, while sporadic Zika cases have been found in Thailand and other Southeast Asian countries, they weren't part of big outbreaks. Only wide serosurveys will tell whether herd immunity from endemic transmission is high all over Africa and Asia, but the lack of intense outbreaks—particularly in Senegal, which is closely connected to the Cape Verde Islands—suggests it is.

In U.S. territories such as Puerto Rico and the Virgin Islands, the epidemic is expected to grow in intensity all summer. Guillain-Barré and other autoimmune diseases like ITP are already on the rise, and on May 13, 2016, Puerto Rico's health department announced that a woman on the island lost a baby that turned out to be microcephalic. Puerto Rico has 3.5 million people, and Dr. Johnny Rullán, the island's former health secretary and now the governor's special assistant for the epidemic, conservatively estimated that every 10,000 circulating infections would trigger one autoimmune reaction, so there could be 350. (French Polynesia's experience would suggest there will be more like 850.)

Ultimately, hundreds or thousands of Guillain-Barré victims needing mechanical ventilation might put a far greater strain on the hemisphere's hospitals than neonates needing intensive care, especially if women decide on their own to hold off pregnancies.

Some doctors in the United States, especially in mosquito-prone areas, are privately suggesting that to patients, and more are saying so publicly. Dr. Edward Goodman, chief epidemiologist at Texas Health Presbyterian, the Dallas hospital made famous two years ago because two of its nurses caught Ebola while caring for a patient, went on television to suggest that women in Dallas consider delaying pregnancy.

At the Zika summit in Atlanta, Dr. Ana Ríus said women in Puerto Rico seemed to be taking her advice: the birthrate was dropping, and the island was on track to have only 28,000 babies in 2016, some 8 percent fewer than in 2015.

Americans' knowledge of the disease is getting more sophisticated, according to the Annenberg Center, which has been doing polls about Zika since February 2016. Early on, half of all Americans worried it would come to their neighborhoods, many thought all mosquitoes had it, and 42 percent thought it was usually fatal. By April, a majority answered that infants and pregnant women are most at risk.

Terrible consequences may *not* come to pass. In the continental United States, Zika may be contained, or may spread far more slowly than anyone fears. Many women may successfully protect themselves through birth control, moving out of danger zones if they can, or minimizing bites.

There certainly will not be the overwhelming flood of microcephalic neonates in American hospitals that there was in Brazil. It will probably not be as bad anywhere else in the hemisphere. Brazil was caught utterly by surprise when thousands of its babies were at or near term. Now many countries' medical establishments are on the alert, doctors are ordering ultrasounds, and many women may choose to terminate. Some will be able to do so legally; some may find another way.

An epidemic averted would be great news. It would be a victory for public health—and for risk communication and freedom of the press, since there is not much medically that can be done as of now.

Of course, if and when that happens, if the early fears are not realized and the dark clouds lift, many people will call it a false alarm. They'll say the media blew it out of proportion.

Fair enough, but the 2009 swine flu scare, which I covered—or contributed to the panic about, if you like—is now widely regarded as a false alarm. It arrived from Mexico in the spring, an unusual time for a flu outbreak, and it too caused a huge number of cases because no one was immune to the new gene mixture. Then it died out in the summer, as flu always does. But a council of White House science advisers predicted that, when it returned in the fall, it would kill up to 90,000 Americans. That document, released late in the day, set off panicky headlines. (Not in my newspaper. I didn't believe the estimate and held off writing until I could reach enough epidemiologists to debunk it. But by then it was too late. *USA Today* and virtually every TV station in the country had featured it prominently.) When the new flu did return, scientists realized it was actually milder than most seasonal flus. Ultimately *fewer* people died of the flu in 2009 than usually did.

But in late 2009, I met Aubrey Opdyke, a 27-year-old former waitress in West Palm Beach, Florida. When she caught the flu in June, she had weighed 135 pounds and had been healthy, except for one condition that put her at high risk in flu season: she was expecting. Nothing scary—she was in the middle of a typical, trouble-free pregnancy.

When I met her in October, she had been home from the hospital only three weeks. She had spent five weeks in a coma, suffered six collapsed lungs and a seizure that nearly killed her. She was still so weak that she needed a walker to get around her living room and could barely lift a one-pound weight during her daily sessions with a physical therapist. From her neck to her ankles, she was all stretch marks: the high-pressure ventilator that had kept her alive during the coma had forced so much air into her tissues that her husband said she'd looked like pictures of 400-pound women.

And she lost the baby. During one of the lung-collapse crises, the infant had to be delivered by Caesarean, and didn't survive. Aubrey was comatose, so she never saw the baby. But her husband named her Parker Christine and let a photographer's charity, "Now I Lay Me Down to Sleep," take black-and-white pictures so that Aubrey would at least have a memory. Her mother bathed Parker and brushed her hair for the photographer—and then they buried her.

"Mild" diseases aren't mild for everyone, and one cynic's false alarm is another mother's disaster. Stay alert. And empathic.

13

Questions and Answers

(The information here is current as of June 1, 2016, and consistent with recommendations from public health authorities at that time. They could change. Please consult reliable websites like cdc.gov.)

Q. How dangerous is Zika?

A. For most people, it's mild. Four out of five people have no symptoms and don't realize they've had it. Most who get it recover in a week to 10 days. There is a small but unpredictable risk of Guillain-Barré paralysis and other dangerous complications. The biggest threat is to unborn children, for whom it can be devastating.

Q. If I'm bitten by a mosquito with Zika, how quickly will I show symptoms?

A. Usually within 3 to 6 days, although it can take as long as 14.

Q. What are the symptoms?

A. The most common are these: A low-grade fever (usually

below 102 degrees). A flat reddish "maculopapular" rash on the trunk—bumpy but not with pustules like those of chicken pox. If you press on it, it disappears—you can even see a clear white handprint. Conjunctivitis—pink or red bloodshot eyes. Pain behind the eyes, especially in bright light. Pains in the back and joints. Not everyone who gets any symptoms gets all of them. The fever and back pain usually precede the rash, but if those are mild, the rash might be your first sign.

Q. If I've been to a Zika transmission area and have been bitten, or if I've noticed these symptoms, what should I do?

A. If you are pregnant, you should see a doctor and get tested as soon as possible. The most accurate tests can be done only in roughly the first 10 days after infection. Tests for antibodies can be done later but take days or weeks more and are less reliable.

Q. What does Zika do to unborn babies?

A. It can cause microcephaly—tiny heads and underdeveloped brains. But it can also kill babies in the womb outright. It may be able to do that at any time during a pregnancy. It may also cause milder damage to nerves in the growing brain that can lead to serious birth defects. Some newborns suffer fatal seizures. Some have spastic or frozen arms and legs. Some cry constantly in a high pitch. Some have difficulty feeding or swallowing. Without intensive care, some die in the first weeks of life.

Q. What is the future for these babies?

A. Most children affected by Zika are still very young, so doctors must guess on the basis of other viruses, like rubella and cytomegalovirus, that attack fetuses. Some babies appear to be blind or deaf. Some may never learn to stand, walk, or control their bowels. Some may have serious learning disabilities from childhood on. And experts fear that some who appear healthy in childhood may develop schizophrenia or bipolar disorder as adults.

Q. What about Guillain-Barré syndrome?

A. Zika can cause Guillain-Barré paralysis, but it's still very rare—in one outbreak the odds of getting it were calculated at about 1 out of every 4,000 Zika cases. You are at some risk of Guillain-Barré at all times; the world baseline rate is about 1 case in 100,000 people per year. It can be triggered by a cold, flu, stomach flu, or surgery. *Campylobacter*, the bacterium found in raw chicken, is a common cause. Vaccines are a rare cause.

Q. Exactly what is Guillain-Barré syndrome?

A. The immune system generates antibodies that attack your own nerve cells. That can cause ascending paralysis, which creeps in from limbs and face to the chest. If it reaches the breathing muscles, a victim can die if he or she is not quickly put on a ventilator. Most people recover within six months to a year. About 15 percent have persistent muscle weakness.

Q. How can I protect myself against Zika?

A. Stay away from areas where mosquitoes are transmitting it. If you cannot, avoid mosquito bites. Do not have unprotected sex with a man who has it or may have it. The CDC advises eight weeks of condom use or abstinence if the man has no symptoms, six months if he has any, and, if you are pregnant, for the length of the pregnancy.

Q. Is it safe for me to travel?

A. It is definitely *not* safe for pregnant women to visit areas with current Zika transmission. Check the websites of the CDC or the WHO for a list of countries with it. But to know whether mosquito activity is intense in a certain region, you may have to do your own research. Anywhere above 6,500 feet is considered safe, but also consider, for example, where you might change planes.

Q. Will it be safe for me to go to the Olympics?

A. The CDC definitely recommends that pregnant women avoid the Olympics. If a husband or boyfriend goes, it recommends no unprotected sex for the rest of the pregnancy. The real infection risk will not be clear until just before the Games, and will depend on how hot and rainy it is. August is Rio's winter, and rainfall is normally low, but "winter" in Rio means temperatures of 65 to 80, and mosquitoes can bite all year around. The Olympic Committee and city government say the venues will be safe, but that promise may prove hollow in a city of 13 million. Rio appears to have had a Zika outbreak in early 2015, and it had a big one from March to May of this year. The WHO has suggested that visitors avoid slums because garbage collects standing water. The CDC has a web page devoted entirely to the Olympics.

Q. What will happen if lots of Americans come back from the Olympics with Zika?

A. Even if that happens, hospitals are unlikely to be overwhelmed because most cases are mild. A bad flu season is probably a bigger threat to the hospital system. But cases of Guillain-Barré may be a problem. Also, travelers returning with Zika may seed outbreaks in their hometowns when local mosquitoes bite them.

Q. What kinds of mosquitoes carry the virus?

A. The most common vector is *Aedes aegypti*, "the yellow fever mosquito." Although it has been found occasionally in 30 states, and in hot, wet years can range as far north as New York City, the threat is expected to be high only in Florida and along the Gulf Coast and in Hawaii. *Aedes albopictus*, "the Asian tiger mosquito," can also carry Zika and is found well north of Chicago and New York in hot summers. But it is not known whether the tiger can spread Zika effectively. Brazil has reported the virus in *Culex* mosquitoes, which are all over the United States. But they've never been shown to transmit it.

Q. How can I tell which mosquitoes are around?

A. Mosquitoes are hard to tell apart. *Aedes* mosquitoes are usually slightly larger than *Culex* or *Anopheles* mosquitoes, which carry different diseases, and they are black with vivid white spots. *Aedes aegypti*, the yellow fever mosquito, has two curved "lyre-shaped" lines on its back. Females lay sticky eggs in clean water—even in pet dishes—and they slip into houses and hide in closets and under beds. They frequently bite ankles and are "sip feeders," biting several people for each blood meal. *Aedes albopictus*, the Asian tiger mosquito, looks similar but can be even bigger and has a white line down its back. It tends to bite and hang on unless squashed.

Q. How can I protect myself against Zika if I live in a transmission area?

A. If you are pregnant, you should avoid bites 24 hours a day. Close or screen all windows and use air-conditioning. Wear long sleeves and pants and repellent with DEET, picaridin, IR3535, oil of lemon eucalyptus, or para-menthane-diol.

Q. Is DEET safe for pregnant women?

A. Yes, according to the CDC, especially in high-risk circumstances like this.

Q. Can I be bitten by a mosquito and still have a healthy baby? What are the chances?

A. Yes. Not every mosquito is an *Aedes*, and not every *Aedes* is infected. By some estimates, even in high-transmission zones only 1 mosquito in 1,000 has the virus. To put this in perspective: Brazil has had over 1,400 confirmed cases of microcephaly and more than 7,000 reported ones. But 3 million babies are normally born in Brazil each year.

Q. If I live in an area where it is being transmitted by mosquitoes, do I also have to protect myself against sexual transmission?

A. Yes. Your husband or boyfriend may get it and pass it on to you even before he has symptoms. You need to avoid unprotected sex—vaginal, anal, and oral.

Q. If I get Zika, and I recover, is it safe for me to have a baby later?

A. Yes, absolutely. A Zika infection is not for life. As with many other rash diseases—chicken pox, smallpox, and measles, for example—getting it once appears to provide lifelong immunity. No one is yet sure whether Zika immunity is lifelong, because it has been studied for only a few years. But doctors believe it is long-lasting.

Q. How long do I have to wait after being in a Zika area before I can have a baby?

A. You are probably safe within about three weeks, but out of caution, the CDC recommends that women wait eight weeks. (They took the estimated safe period and nearly tripled it.)

Q. Should I delay having a baby this year?

A. That is a difficult question that every woman has to answer for herself. Some leading doctors think women in areas where there is Zika transmission now—or may be soon—would be wise to delay pregnancy if they can. It is virtually impossible to protect yourself against mosquito bites for nine months. Epidemics are usually fiercest in their first year. And in a few years, a vaccine may be available.

Q. What about my husband? Can he give me Zika?

A. Yes. Doctors are still learning more about this risk. If a man has had no symptoms and no test after returning from an area with transmission, the CDC recommends avoiding any contact with his semen for eight weeks—meaning either abstain from vaginal, oral, and anal sex or use condoms. If he had symptoms or a positive Zika test, they recommend avoiding contact with his semen for six months.

Q. Six months? Why so long?

A. Live Zika virus has been found in semen as long as two months after symptoms disappear. Pregnant women should avoid contact with his semen for the entire pregnancy.

Q. Can my husband get Zika without ever knowing it?

A. Definitely. Eighty percent of all people infected with Zika never display symptoms.

Q. If my husband has been to a place with Zika, what are the chances that he got it?

A. Unfortunately, that is just not knowable. The risk varies not just by country but by region as well. A buggy lowland area of Mexico can be dangerous, while Mexico City may be perfectly safe because it's located too high for mosquitoes.

Q. Can my husband/boyfriend get it from sex with another woman who has Zika?

A. Probably not. As of this writing, there have been no documented cases of female-to-male human transmission. So if he says, "I swear, dear, I got it from a mosquito bite," he's probably telling the truth.

Q. Can my husband/boyfriend get it from sex with another man who has Zika?

A. Yes. But if your husband or boyfriend is bisexual, you may face other risks, including HIV, syphilis, gonorrhea, and chlamydia, which are more common among gay and bisexual men.

Q. Can my husband or boyfriend give me Zika even though he never felt sick?

A. Possibly. Men have definitely transmitted Zika to their wives *before* falling ill. Whether a man can have no symptoms at all and still transmit the infection is unknown.

Q. Should men who have returned from a Zika area be sperm donors?

A. There have been no known cases of transmission that way, but since it's a sexually transmitted disease, sperm banks should take the same precautions against it that they do against other STI's.

Q. If I get inseminated with sperm from a sperm bank, what are the chances it has Zika virus in it?

A. This is a new area. But in theory at least, the chances should be low. Sperm banks should screen donors and sperm should be tested for virus. However, no test is perfect.

Q. Can I get Zika from anal sex, oral sex, or any other form of sex?

A. Definitely from anal sex. Transmission by oral sex is suspected but not proven. The virus has been found at high levels in semen, blood, and urine, and at low levels in saliva. Transmission by kissing has not been documented and is thought to be unlikely. Contact between infectious fluids and mucus membranes like the insides of vaginas, rectums, and mouths, or with the eyes or nose, is not considered safe. Contact with hands, breasts, or any part of the body covered by intact skin probably is.

Q. I'm a gay man. Can I get Zika from anal sex from my boyfriend?

A. Absolutely, yes. It has happened.

Q. If I'm pregnant, and have had Zika, how soon will I be able to tell whether my baby has been hurt?

A. If you had a positive Zika test, your doctor should schedule ultrasounds and MRIs as often as every three weeks. They may suggest amniocentesis, to look for virus in the fluid around the baby. No one knows how long from the date of infection it takes for damage to show up on an ultrasound, but changes indicating brain damage have been detected as early as week 19 of pregnancy, in the second trimester.

Q. If I have had Zika, what are the chances that my baby has been hurt?

A. No one knows. The majority of babies whose mothers had Zika appear normal at birth. Very early studies suggested the chances of brain damage are somewhere between 1 percent and 29 percent, which is a very wide margin. A CDC-sponsored study published May 25 found the risk of microcephaly to be as high as 13 percent. It did not make estimates for other types of fetal brain damage. Bigger studies that should answer the question more accurately are now underway.

Q. If I have Zika, can I safely breastfeed my child?

A. Although the virus has been found in breast milk, breastfeeding has has never been proven to transmit it, so the WHO and CDC recommend that women with Zika continue to breastfeed. They believe the benefits greatly outweigh the risks.

Q. Is there a Zika vaccine? Will there be?

A. No, not yet. Nearly 20 laboratories are working on them, but lengthy testing is mandatory. The most optimistic scenarios hope for one by late 2018. Sometime before 2021 is considered more realistic. Some pessimists fear the risk of triggering Guillain-Barré paralysis will make a vaccine impossible.

Notes

CHAPTER 1

15 "Zika doesn't worry us": Donald G. McNeil Jr., Simon Romero, and Sabrina Tavernise, "How a Medical Mystery in Brazil Led Doctors to Zika," *New York Times*, Feb. 6, 2016, http://www.nytimes.com/2016/02/07/health/zika-virus-brazil-how-it-spread-explained.html.

CHAPTER 2

18 "Ziika"—the spelling was shortened: Josh Kron, "In a Remote Ugandan Lab, Encounters with the Zika Virus and Mosquitoes Decades Ago," *New York Times*, April 5, 2016, http://www.nytimes.com/2016/04/06/world/africa/uganda-zika-forest-mosquitoes.html.

19 On April 19, 1947: Jon Cohen, "Zika's Long Strange Trip into the Limelight," *Science*, Feb. 8, 2016, http://www.sciencemag.org/news/2016/02/zika-s-long-strange-trip-limelight.

20 a "filterable, transmissible agent": G. W. A. Dick, S. F. Kitchen, and A. J. Haddow, "Zika Virus (I). Isolations and Serological Specificity" *Transactions of the Royal Society of Tropical Medicine and Hygiene*, Sept. 1952, p. 509, http://trstmh.oxfordjournals.org/content/46/5/509.full.pdf+html.

25 "an African female aged 10": F. N. Macnamara, "Zika Virus: A Report on Three Cases of Human Infection during an Epidemic of Jaundice in Nigeria," *Transactions of the Royal Society of Tropical Medicine and Hygiene*, March 1954, p. 139, http://trstmh.oxfordjournals.org/content/48/2/139.full.pdf+html?sid=864a5093-2f7f-4f0b-ba6d-30030da79028.

27 After marking the spot: W. G. C. Bearcroft, "Zika Virus Infection Experimentally Induced in a Human Volunteer," *Transactions of the Royal Society of Tropical Medicine and Hygiene*, Sept. 1956, p. 442, http://trstmh.oxfordjournals.org/content/50/5/442.full.pdf+html?sid=ad409066-59d5-4ef4-a3a7-29dd4a149ae1.

28 In 1964, another researcher: D. I. H. Simpson, "Zika Virus Infection in Man," *Transactions of the Royal Society of Tropical Medicine and Hygiene*, July 1964, p. 335, http://linkinghub.elsevier.com/retrieve/pii/0035920364902007?showall=true.

32 In March 2016, researchers: Shannan L. Rossi et al., "Characterization of a Novel Murine Model to Study Zika Virus," *American Journal of Tropical Medicine and Hygiene*, March 28, 2016, doi:10.4269/ajtmh.16-0111.

CHAPTER 3

37 The first time Zika was noticed: Mark R. Duffy et al., "Zika Virus Outbreak on Yap Island, Federated States of Micronesia," *New England Journal of Medicine*, June 11, 2009, doi:10.1056/NEJMoa0805715.

38 Dengue kept looking: Austin Ramzy, "Experts Study Zika's Path from First Outbreak in Pacific," *New York Times*, Feb. 10, 2016, http://www.nytimes.com/2016/02/11/world/asia/zika-virus-yap-island.html.

39 "We worked through hot": Tai-Ho Chen et al., "Zika Virus Outbreak—Yap, Micronesia, June 2007," *EIS e-Bulletin*, March 2008.

40 "Our health care system": Reuters Health E-Line, "Little Known Virus Causes Outbreak in Pacific Isles," Reuters, July 10,

2007, http://www.reuters.com/article/us-virus-mosquito-zika-idUSHKG4179220070710.

41 On October 7, 2013: Henri-Pierre Mallet et al., "Epidémie de virus Zika en Polynésie française" (PowerPoint presentation at interregional conference on public health surveillance, Fort de France, Martinique, Nov. 5–7, 2015).

41 "Tahiti is a small island": Jason Beaubien, "Zika in French Polynesia: It Struck Hard in 2013, Then Disappeared," NPR News, Feb. 9, 2016, http://www.npr.org/sections/goatsandsoda/2016/02/09/466152313/zika-in-french-polynesia-it-struck-hard-in-2013-then-disappeared.

41 The first household: Van-Mai Cao-Lormeau et al., "Zika Virus, French Polynesia, South Pacific, 2013" (letter), *Emerging Infectious Diseases*, June 1, 2014, doi:10.3201/eid2006.140138.

42 The first was a woman: E. Oehler et al., "Zika Virus Infection Complicated by Guillain-Barré Syndrome—Case Report, French Polynesia, December 2013," *Eurosurveillance*, March 6, 2014, doi:http://dx.doi.org/10.2807/1560-7917.ES2014.19.9.20720.

43 "Up till then, everyone": Damien Mascret, "Interview with Dr. Sandrine Mons," *Le Figaro*, Feb. 18, 2016, http://sante.lefigaro.fr/actualite/2016/02/18/24642-tahiti-zika-etait-considere-comme-benin.

44 Four cases of "immune": "Surveillance de la dengue et du zika en Polynésie française," *Centre d'Hygiène et du Salubrité Publique*, Feb. 7, 2014, http://www.hygiene-publique.gov.pf/IMG/pdf/bulletin_dengue_07-02-14.pdf.

44 The worst-off was Larry Ly: Karen Weintraub, "Scientists Link Zika Firmly to Paralysis, as Patients in Tahiti Know Too Well," *STAT News*, Feb. 29, 2016, https://www.statnews.com/2016/02/29/zika-guillain-barre-tahiti/.

44 As fear of the disease: "Epidémie de Zika: La ministre de la santé au chevet des malades," *Tahiti Infos*, Feb. 22, 2014, http://www.tahiti-infos.com/forum/Epidemie-de-Zika-La-ministre-de-la-sante-au-chevet-des-malades_m195232.html.

45 Finally, the French high commissioner: "Les maires résistants aux pulvérizations d'insecticide rappelés à l'ordre," *Tahiti Infos*, Feb. 19, 2014, http://www.tahiti-infos.com/L-epidemie-de-zika-recule-mais-la-situation-sanitaire-reste-tendue_a95004.html.

45 By April 2014, when: Henri-Pierre Mallet, "Emergence du virus Zika en Polynésie française" (PowerPoint presentation at national infectious disease conference, Bordeaux, France, June 11–13, 2014).

46 The Guillain-Barré "attack": Van-Mai Cao-Lormeau et al., "Guillain-Barré Syndrome Outbreak Associated with Zika Virus Infection in French Polynesia: A Case-Control Study," *Lancet*, Feb. 29, 2016, http://dx.doi.org/10.1016/S0140-6736(16)00562-6.

46 In an interview: Gwendoline Dos Santos and Frédéric Lewino, "Scandal: Health Authorities Ignore a Leading Zika Specialist," *Le Point*, Feb. 3, 2016, http://www.lepoint.fr/sante/zika-la-propagation-de-l-epidemie-etait-previsible-de puis-2014-03-02-2016-2014974_40.php.

46 In 2015, when France's High Council: Dos Santos and Lewino, "Scandal."

47 From Tahiti, the virus: "L'épidémie de zika en phase terminale en Polynésie française," *Tahiti Infos*, April 9, 2014, http://www.tahiti-infos.com/L-epidemie-de-zika-en-phase-terminale-en-Polynesie-francaise_a98643.html.

48 Dr. Musso then: Didier Musso, "Zika Virus Transmission from French Polynesia to Brazil," *Emerging Infectious Diseases*, Oct. 2015, http://dx.doi.org/10.3201/eid2110.151125.

48 But in March 2016: Nuno Rodrigues Faria et al., "Zika Virus in the Americas: Early Epidemiological and Genetic Findings," *Science*, March 24, 2016, http://science.sciencemag.org/content/early/2016/03/23/science.aaf5036.

48 In April, researchers: John Lednicky et al., "Zika Virus Outbreak in Haiti in 2014: Molecular and Clinical Data," *PLOS Neglected Tropical Diseases*, April 25, 2016, http://dx.doi.org/10.1371/journal.pntd.0004687.

49 This was not the first: "L'origine de l'épidémie de zika en Polynésie restera un mystère," *Tahiti Infos*, Dec. 9, 2013, http://www.tahiti-infos.com/Dengue-Zika-la-chasse-aux-moustiques-doit-etre-totale_a89989.html.

CHAPTER 4

52 Hospital hallways: Donald G. McNeil Jr., Simon Romero, and Sabrina Tavernise, "How a Medical Mystery in Brazil Led Doctors to Zika," *New York Times*, Feb. 6, 2016, http://www.nytimes.com/2016/02/07/health/zika-virus-brazil-how-it-spread-explained.html.

53 As well they might: J. P. Messina et al., "Mapping Global Environmental Suitability for Zika Virus," *eLife Sciences*, Jan. 18, 2016, http://simonhay.well.ox.ac.uk/uploads/publications/309/Messina_MappingGlobalEnvSuitZikaVirus_2016_provis_SI.pdf.

54 After that, the greatest: Van-Mai Cao-Lormeau et al., "Guillain-Barré Syndrome Outbreak Associated with Zika Virus Infection in French Polynesia: A Case-control Study," *Lancet*, Feb. 29, 2016, http://dx.doi.org/10.1016/S0140-6736(16)00562-6.

CHAPTER 5

59 He forwarded me: S. C. Weaver and W. K. Reisen, "Present and Future Arboviral Threats," *Antiviral Research*, Oct. 24, 2009, doi:10.1016/j.antiviral.2009.10.008.

61 On Google News: Shasta Darlington, "Brazil Warns against Pregnancy Due to Spreading Virus," CNN News, Dec. 24, 2015, http://www.cnn.com/2015/12/23/health/brazil-zika-pregnancy-warning/.

62 A few years earlier: Donald G. McNeil Jr., "Fast New Test Could Find Leprosy before Damage Is Lasting," *New York Times*, Feb. 19, 2013, http://www.nytimes.com/2013/02/20/health/fast-new-test-could-help-nip-leprosy-in-the-bud.html.

63 I wrote a brief: Donald G. McNeil Jr., "Zika Virus, a Mosquito-Borne Infection, May Threaten Brazil's Newborns," *New York Times*, Dec. 28, 2015, http://www.nytimes.com/2015/12/29/health/zika-virus-brazil-mosquito-brain-damage.html.

66 Simon's story arrived: Simon Romero, "Alarm Spreads in Brazil over a Virus and a Surge in Malformed Infants," *New York Times*, Dec. 30, 2015, http://www.nytimes.com/2015/12/31/world/americas/alarm-spreads-in-brazil-over-a-virus-and-a-surge-in-malformed-infants.html.

67 On January 4, 2016: Donald G. McNeil Jr., "U.S. Becomes More Vulnerable to Tropical Diseases Like Zika," *New York Times*, Jan. 4, 2016, http://www.nytimes.com/2016/01/05/health/us-becomes-more-vulnerable-to-tropical-diseases-like-zika.html.

67 Then, on January 4, Puerto Rico: Lisa Schnirring, "Puerto Rico Reports First Locally Acquired Zika Virus Case," *CIDRAP News*, Jan. 4, 2016, http://www.cidrap.umn.edu/news-perspective/2016/01/puerto-rico-reports-first-locally-acquired-zika-virus-case.

68 Actually, I learned: "First Case of Zika Virus Reported in Puerto Rico," *Centers for Disease Control and Prevention*, Dec. 31, 2015, http://www.cdc.gov/media/releases/2015/s1231-zika.html.

68 until an Associated Press: "Puerto Rico Reports Its First Mosquito-Borne Zika Case," Associated Press, Jan. 4, 2016, http://wgntv.com/2016/01/04/puerto-rico-reports-its-first-mosquito-borne-zika-case/.

72 On January 13, I got: Lavinia Schuler-Faccini et al., "Possible Association between Zika Virus Infection and Microcephaly—Brazil, 2015," *Morbidity and Mortality Weekly Report*, Jan. 22, 2016, http://dx.doi.org/10.15585/mmwr.mm6503e2.

72 This was a *second*: "Zika Virus Epidemic in the Americas: Potential Association with Microcephaly and Guillain-Barré Syndrome," *European Centre for Disease Prevention and Control*, Dec. 10, 2015, p. 7, http://ecdc.europa.eu/en/publications/Publications/zika-virus-americas-association-with-microcephaly-rapid-risk-assessment.pdf.

75 The agency issued: Donald G. McNeil Jr., "To Protect against Zika Virus, Pregnant Women Are Warned about Latin American Trips," *New York Times*, Jan. 15, 2016, http://www.nytimes.com/2016/01/16/health/zika-virus-cdc-pregnant-women-travel-warning.html.

76 That must really have upset: Simon Romero, "Zika Warning Spotlights Latin America's Fight against Mosquito-Borne Diseases," *New York Times*, Jan. 17, 2016, http://www.nytimes.com/2016/01/18/world/americas/brazil-zika-warning-rio-games.html.

76 The first American baby: Donald G. McNeil Jr., "Hawaii Baby with Brain Damage Is First U.S. Case Tied to Zika Virus," *New York Times*, Jan. 16, 2016, http://www.nytimes.com/2016/01/17/health/hawaii-reports-baby-born-with-brain-damage-linked-to-zika-virus.html.

CHAPTER 6

77 People all over: Andrew Jacobs, "Brazilians Shrug Off Zika Fears to Revel in Carnival Fun," *New York Times*, Feb. 10, 2016, http://www.nytimes.com/2016/02/11/world/americas/brazil-zika-virus-carnival.html.

77 Up to then, the WHO: Editorial, "Zika Virus Requires an Urgent Response," *New York Times*, Jan. 28, 2016, http://www.nytimes.com/2016/01/28/opinion/zika-virus-requires-an-urgent-response.html.

78 two weeks later: Sabrina Tavernise and Donald G. McNeil Jr., "Zika Virus a Global Health Emergency, W.H.O. Says," *New York Times*, Feb. 1, 2016, http://www.nytimes.com/2016/02/02/health/zika-virus-world-health-organization.html.

81 In fact, within a week: "Statement on Data Sharing in Public Health Emergencies," *PLOS in the News*, Feb. 10, 2016, http://blogs.plos.org/plos/2016/02/statement-on-data-sharing-in-public-health-emergencies/.

83 My colleague Catherine: Catherine Saint Louis, "Microceph-

aly, Spotlighted by Zika Virus, Has Long Afflicted and Mystified," *New York Times*, Jan. 31, 2016, http://www.nytimes.com/2016/02/01/health/microcephaly-spotlighted-by-zika-virus-has-long-afflicted-and-mystified.html.

83 The BBC did: Graham Satchell, "'He Is Enjoying Life': Living with Microcephaly in the UK," BBC News, Feb. 5, 2016, http://www.bbc.com/news/health-35500306.

83 French Polynesian scientists: Dr. Simon Cauchemez et al., "Association between Zika Virus and Microcephaly in French Polynesia, 2013–15: A Retrospective Study," *Lancet*, March 15, 2016, http://dx.doi.org/10.1016/S0140-6736(16)00651-6.

84 Sabrina Tavernise interviewed: Donald G. McNeil Jr., Simon Romero, and Sabrina Tavernise, "How a Medical Mystery in Brazil Led Doctors to Zika," *New York Times*, Feb. 6, 2016, http://www.nytimes.com/2016/02/07/health/zika-virus-brazil-how-it-spread-explained.html.

84 a story about mosquito control: Sabrina Tavernise, "Prepare for Guerrilla Warfare with Zika-Carrying Mosquitoes, Experts Warn," *New York Times*, Feb. 12, 2016, http://www.nytimes.com/2016/02/13/health/prepare-for-guerrilla-warfare-with-zika-carrying-mosquitoes-experts-warn.html.

84 Simon did one: Simon Romero and Donald G. McNeil Jr., "Zika Virus May Be Linked to Surge in Rare Syndrome in Brazil," *New York Times*, Jan. 22, 2016, http://www.nytimes.com/2016/01/22/world/americas/zika-virus-may-be-linked-to-surge-in-rare-syndrome-in-brazil.html.

84 The first American to die: Donald G. McNeil Jr. and Daniel Victor, "First U.S. Death Tied to Zika Is Reported in Puerto Rico," *New York Times*, April 29, 2016, http://www.nytimes.com/2016/04/30/health/zika-virus-first-death-in-us-puerto-rico.html.

86 And there was no way: Joari De Miranda et al., "Induction of Toll-Like Receptor 3-Mediated Immunity during Gestation Inhibits Cortical Neurogenesis and Causes Behavioral Disturbances," *mBio*, Oct. 5, 2010, doi:10.1128/mBio.00176-10.

87 In 1988, Finnish: Sarnoff A. Mednick et al., "Adult Schizophrenia Following Prenatal Exposure to an Influenza Epidemic," *Archives of General Psychiatry* 45, no. 2 (1988): 189, doi:10.1001/archpsyc.1988.01800260109013.

87 And a pioneer: E. Fuller Torrey, interview with the author, Feb. 2016.

88 Whether psychiatric problems: Donald G. McNeil Jr., "Zika May Increase Risk of Mental Illness, Researchers Say," *New York Times*, Feb. 18, 2016, http://www.nytimes.com/2016/02/23/health/zika-may-increase-risk-of-mental-illness-researchers-say.html.

88 Eventually, scientists from: Cauchemez et al., "Association."

CHAPTER 7

94 The only obvious: Brian D. Foy et al., "Probable Non-Vector-Borne Transmission of Zika Virus, Colorado, USA," *Emerging Infectious Diseases*, May 17, 2011, p. 880, doi:10.3201/eid1705.101939.

94 The paper described: Martin Enserink, "Sex after a Field Trip Yields Scientific First," *Science*, April 6, 2011, http://www.sciencemag.org/news/2011/04/sex-after-field-trip-yields-scientific-first.

95 It was in a relatively minor: Didier Musso et al., "Potential Sexual Transmission of Zika Virus," *Emerging Infectious Diseases*, Feb. 2015, p. 359, doi:10.3201/eid2102.141363.

95 On January 25, 2016: Donald G. McNeil Jr., "Zika Virus: Two Cases Suggest It Could Be Spread through Sex," *New York Times*, Jan. 25, 2016, http://www.nytimes.com/2016/01/26/health/two-cases-suggest-zika-virus-could-be-spread-through-sex.html.

98 In retrospect, part of: D. Trew Deckard et al., "Male-to-Male Sexual Transmission of Zika Virus—Texas, January 2016," *Morbidity and Mortality Weekly Report*, April 15, 2016, http://www.cdc.gov/mmwr/volumes/65/wr/mm6514a3.htm.

100 The eyes are privileged: Denise Grady, "After Nearly Claiming His Life, Ebola Lurked in a Doctor's Eye," *New York Times*, May

7, 2015, http://www.nytimes.com/2015/05/08/health/weeks-after-his-recovery-ebola-lurked-in-a-doctors-eye.html.

CHAPTER 8

102 a nurse-practitioner: Dyan J. Summers et al., "Zika Virus in an American Recreational Traveler," *Journal of Travel Medicine*, May 21, 2015, doi:10.1111/jtm.12208.

106 It was a good call: Henri-Pierre Mallet, "Emergence du virus Zika en Polynésie française" (PowerPoint presentation at national infectious disease conference, Bordeaux, France, June 11–13, 2014).

CHAPTER 9

112 A 2008 study: Celia W. Dugger, "Study Cites Toll of AIDS Policy in South Africa," *New York Times*, Nov. 25, 2008, http://www.nytimes.com/2008/11/26/world/africa/26aids.html.

115 The rumor about a larvicide: Production Team REDUAS, "Report from Physicians in the Crop-Sprayed Towns regarding Dengue-Zika, Microcephaly, and Massive Spraying with Chemical Poisons," *Red Universitaria de Ambiente y Salud*, Feb. 9, 2016, http://www.reduas.com.ar/informe-de-medicos-de-pueblos-fumigados-sobre-dengue-zika-y-fumigaciones-con-venenos-quimicos/.

119 I dug through: "Microcephaly—Brazil," *WHO Disease Outbreak News*, Dec. 15, 2015. http://www.who.int/csr/don/15-december-2015-microcephaly-brazil/en/.

120 Then, on February 3: Vinod Sreeharsha, "Birth Defects in Brazil May Be Overreported amid Zika Fears," *New York Times*, Feb. 3, 2016, http://www.nytimes.com/2016/02/04/world/americas/birth-defects-in-brazil-may-be-over-reported-amid-zika-fears.html.

CHAPTER 10

125 Scientists in Latin America: Donald G. McNeil Jr., "Proof of Zika's Role in Birth Defects Still Months Away, W.H.O. Says," *New York Times*, Feb. 19, 2016, http://www.nytimes.com/2016/02/20/health/zika-virus-microcephaly-birth-defects-proof-who.html.

126 "Based on observational": "Zika Situation Report," *WHO Emergencies*, March 31, 2016, http://www.who.int/emergencies/zika-virus/situation-report/31-march-2016/en/.

126 Then on April 13: Pam Belluck and Donald G. McNeil Jr., "Zika Virus Causes Birth Defects, Health Officials Confirm," *New York Times*, April 13, 2016, http://www.nytimes.com/2016/04/14/health/zika-virus-causes-birth-defects-cdc.html.

127 A series of small studies: Donald G. McNeil Jr., "6 Reasons to Think the Zika Virus Causes Microcephaly," *New York Times*, updated May 3, 2016, http://www.nytimes.com/interactive/2016/04/01/health/02zika-microcephaly.html.

127 Separate teams of doctors: Jernej Mlakar, MD, et al., "Brief Report: Zika Virus Associated with Microcephaly," *New England Journal of Medicine*, Feb. 10, 2016, doi:10.1056/NEJMoa1600651.

127 One particularly grim: Rita W. Driggers et al., "Zika Virus Infection with Prolonged Maternal Viremia and Fetal Brain Abnormalities," *New England Journal of Medicine*, March 30, 2016, doi:10.1056/NEJMoa1601824.

128 There was also "biological": Hengli Tang et al., "Zika Virus Infects Human Cortical Neural Progenitors and Attenuates Their Growth," *Cell Stem Cell*, March 4, 2016, http://dx.doi.org/10.1016/j.stem.2016.02.016.

128 But the most convincing: Patrícia Brasil et al., "Zika Virus Infection in Pregnant Women in Rio de Janeiro—Preliminary Report," *New England Journal of Medicine*, March 4, 2016, doi:10.1056/NEJMoa1602412.

130 A study done by the CDC: Michael A. Johansson et al. "Zika

and the Risk of Microcephaly," *New England Journal of Medicine*, online, May 25, 2016, doi: 10.1056/NEJMp1605367.

130 The stack of evidence: Sonja A. Rasmussen, MD, et al., "Special Report: Zika Virus and Birth Defects—Reviewing the Evidence for Causality," *New England Journal of Medicine*, April 13, 2016, doi:10.1056/NEJMsr1604338.

CHAPTER 11

136 Then El Salvador proposed: Azam Ahmed, "El Salvador's Advice on Zika Virus: Don't Have Babies," *New York Times*, Jan. 25, 2016, http://www.nytimes.com/2016/01/26/world/americas/el-salvadors-advice-on-zika-dont-have-babies.html.

136 They began with: Anastasia Moloney, "Advice to Delay Pregnancy Due to Zika Virus Is Naïve, Activists Say," Reuters, Jan. 22, 2016, http://www.reuters.com/article/us-americas-health-zika-idUSKCN0V100H.

137 Sara Garcia: Moloney, "Advice."

137 American activists: Moloney, "Advice."

138 NPR's *Morning Edition*: Renee Montagne, "Is It Realistic to Recommend Delaying Pregnancy during Zika Outbreak?," NPR's *Morning Edition*, aired Jan. 27, 2016, http://www.npr.org/sections/goatsandsoda/2016/01/27/464533090/is-it-realistic-to-recommend-delaying-pregnancy-during-zika-outbreak.

138 Almost immediately, *Time*: Charlotte Alter, "Why Latin American Women Can't Follow the Zika Advice to Avoid Pregnancy," *Time*, Jan. 28, 2016, http://time.com/4197318/zika-virus-latin-america-avoid-pregnancy/.

138 Emma Saloranta, a founder: Emma Saloranta, "Zika Virus and the Hypocrisy of Telling Women to Delay Pregnancy," *Huffington Post*, Jan. 29, 2016, http://www.huffingtonpost.com/emma-saloranta/zika-virus-and-the-hypocrisy-of-telling-women-to-delay-pregnancy_b_9090476.html.

143 I also learned: Eliza Berman, "How a German Measles Epidemic

Stoked the Abortion Debate in 1965," *Time*, Feb. 2, 2015, http://time.com/3692652/measles-abortion-1960s/.

144 On February 5: Donald G. McNeil Jr., "Growing Support among Experts for Zika Advice to Delay Pregnancy," *New York Times*, Feb. 5, 2016, http://www.nytimes.com/2016/02/09/health/zika-virus-women-pregnancy.html.

149 Pope Francis had: Simon Romero and Jim Yardley, "Francis Says Contraception Can Be Used to Slow Zika," *New York Times*, Feb. 18, 2016, http://www.nytimes.com/2016/02/19/world/americas/francis-says-contraception-can-be-used-to-slow-zika.html.

149 On March 8: Donald G. McNeil Jr., "W.H.O. Advises Pregnant Women to Avoid Areas Where Zika Is Spreading," *New York Times*, March 8, 2016, http://www.nytimes.com/2016/03/09/health/zika-virus-pregant-women-travel.html.

150 I emailed him: Sabrina Tavernise, "W.H.O. Recommends Contraception in Countries with Zika Virus," *New York Times*, Feb. 18, 2016, http://www.nytimes.com/2016/02/19/health/zika-virus-birth-control-contraception-who.html.

151 On March 25: Sabrina Tavernise, "C.D.C. Offers Guidelines for Delaying Pregnancy after Zika Exposure," *New York Times*, March 25, 2016, http://www.nytimes.com/2016/03/26/health/zika-virus-pregnancy-cdc-waiting-period.html.

156 On April 14: Donald G. McNeil Jr., "Health Officials Split over Advice on Pregancy in Zika Areas," *New York Times*, April 14, 2016, http://www.nytimes.com/2016/04/15/health/zika-virus-pregnancy-delay-birth-defects-cdc.html.

157 Houston had just: Peter J. Hotez, "Zika Is Coming," *New York Times*, April 8, 2016, http://www.nytimes.com/2016/04/09/opinion/zika-is-coming.html.

157 CBS News interviewed him: Jonathan LaPook, "Houston Authorities Bracing for Zika Virus Appearance," *CBS Evening News*, aired April 22, 2016, http://www.cbsnews.com/news/houston-texas-authorities-bracing-for-zika-virus-arrival/.

157 He had said: Hari Sreenivasan, "Is a Perfect Storm of Zika Virus

Conditions Coming to the Gulf Coast?," *PBS NewsHour*, aired April 18, 2016, http://www.pbs.org/newshour/bb/is-a-perfect-storm-of-zika-virus-conditions-coming-to-the-gulf-coast/.

157 PAHO was leaning: McNeil, "Health Officials."

CHAPTER 12

160 The other reason: "Locally Acquired Dengue—Key West, Florida, 2009–2010," *CDC Morbidity and Mortality Weekly Report*, May 21, 2010, http://www.cdc.gov/mmwr/preview/mmwrhtml/mm5919a1.htm.

160 In 2009, it took only: Amesh A. Adalja et al., "Lessons Learned during Dengue Outbreaks in the United States, 2001–2011," *Emerging Infectious Diseases*, April 2012, doi:10.3201/eid1804.110968.

160 Even more impressive: Charles Simmins, "Rochester Patient Links to Dengue Fever Outbreak on Florida," *Examiner.com*, May 22, 2010, http://www.examiner.com/article/rochester-patient-links-to-dengue-fever-outbreak-florida.

161 Zika's spread: "West Nile Virus in the United States," *Wikipedia*, last modified Dec. 11, 2015, https://en.wikipedia.org/wiki/West_Nile_virus_in_the_United_States.

161 West Nile is: CDC>West Nile Virus Home>Statistics and Maps, last updated June 9, 2015, http://www.cdc.gov/westnile/statsmaps/index.html.

162 But there is no: "Estimated Range of *Aedes aegypti* and *Aedes albopictus* in the United States, 2016 Maps," CDC>Zika Virus Home>Information for Specific Groups>Vector Surveillance & Control, updated April 26, 2016, http://www.cdc.gov/zika/vector/range.html.

163 A majority of Americans: "A Majority of People Say They Would Get a Zika Vaccine If It Were Available," Annenberg Public Policy Center of the University of Pennsylvania, May 12, 2016, http://www.annenbergpublicpolicycenter.org/a-majority-of-people-say-theyd-get-a-zika-vaccine-if-it-were-available/.

163 At least 18: Katie Thomas, "Vaccine for Zika Virus May Be Years Away, Experts Warn," *New York Times*, Jan. 29, 2016, http://www.nytimes.com/2016/01/30/business/vaccine-for-zika-virus-may-be-years-away-disease-experts-warn.html.

164 The National Institutes of: Anthony S. Fauci, MD, and David M. Morens, MD, "Zika Virus in the Americas—Yet Another Arbovirus Threat," *New England Journal of Medicine*, Jan. 13, 2016, doi:10.1056/NEJMp1600297.

165 Meanwhile, safety testing: Rafick-Pierre Sekaly, "The Failed HIV Merck Vaccine Study: A Step Back or a Launching Point for Future Vaccine Development?," *Journal of Experimental Medicine*, Jan. 14, 2008, doi:10.1084/jem.20072681.

167 Viruses, on the other hand: "Faster, Easier Cures for Hepatitis C," *FDA's Consumer Updates*, July 28, 2014, http://www.fda.gov/ForConsumers/ConsumerUpdates/ucm405642.htm.

168 There is no: Devika Sirohi et al., "The 3.8 Å resolution Cryo-EM Structure of Zika Virus," Science, April 22, 2016, doi:10.1126/science.aaf5316.

168 One paper published: Eloise Mastrangelo et al., "Ivermectin Is a Potent Inhibitor of Flavivirus Replication Specifically Targeting NS3 Helicase Activity: New Prospects of an Old Drug," *Journal of Antimicrobial Chemotherapy*, April 25, 2012, doi:10.1093/jac/dks147.

168 Other papers say: Joanna Zmurko et al., "The Viral Polymerase Inhibitor 7-Deaza-2'-C-Methyladenosine Is a Potent Inhibitor of In Vitro Zika Virus Replication and Delays Disease Progression in a Robust Mouse Model Infection Model," *PLOS Neglected Tropical Diseases*, May 10, 2016, doi:10.1371/journal.pntd.0004695; Marcos Carvalho Borges et al., "Chloroquine Use Improves Dengue-Related Symptoms," *Memórias do Institutio Oswaldo Cruz*, Aug. 2013, doi:10.1590/0074-0276108052013010.

168 Antibodies are expensive: Andrew Pollack, "Experimental Drug Would Help Fight Ebola If Supply Increases, Study Finds," *New York Times*, Aug. 29, 2014, http://www.nytimes.com/2014/08/30/world/africa/study-says-zmapp-works-against-ebola-but-making-it-takes-time.html.

169 Immunoglobulin: E. Oehler et al., "Zika Virus Infection Complicated by Guillain-Barré Syndrome—Case Report, French Polynesia, December 2013," *Eurosurveillance*, March 6, 2014, doi:http://dx.doi.org/10.2807/1560-7917.ES2014.19.9.20720.

169 CDC guidelines: "Clinical Guidance: Pregnant Women and Women of Reproductive Age," CDC>Zika Virus Home>For Health, http://www.cdc.gov/zika/hc-providers/clinical-guidance.html.

169 When microcephaly began: Catherine Saint Louis, "Microcephaly, Spotlighted by Zika Virus, Has Long Afflicted and Mystified," *New York Times*, Jan. 31, 2016, http://www.nytimes.com/2016/02/01/health/microcephaly-spotlighted-by-zika-virus-has-long-afflicted-and-mystified.html.

169 That time has been reduced: Rita W. Driggers et al., "Zika Virus Infection with Prolonged Maternal Viremia and Fetal Brain Abnormalities," *New England Journal of Medicine*, March 30, 2016, doi:10.1056/NEJMoa1601824.

169 Because no one: "PAHO Statement on Zika Virus Transmission and Prevention," *Pan American Health Organization*, Feb. 2, 2016, http://www.paho.org/hq/index.php?option=com_content&view=article&id=11605&Itemid=41716&lang=en.

170 It also reached: "Zika Situation Report 19 May 2016," *World Health Organization Emergencies*, May 19, 2016, http://www.who.int/emergencies/zika-virus/situation-report/19-may-2016/en/.

170 The initial worries: Donald G. McNeil Jr., "C.D.C. Is Monitoring 279 Pregnant Women with Possible Zika Virus Infections," *New York Times*, May 20, 2016, http://www.nytimes.com/2016/05/21/health/pregnant-women-zika-virus-cdc.html.

171 At the Zika summit: Donald G. McNeil Jr., "Health Officials Split over Advice on Pregancy in Zika Areas," *New York Times*, April 14, 2016, http://www.nytimes.com/2016/04/15/health/zika-virus-pregnancy-delay-birth-defects-cdc.html.

171 Americans' knowledge: "Half of Americans Concerned Zika Will Spread to Their Neighborhoods," Feb. 23, 1016, http://www

.annenbergpublicpolicycenter.org/half-of-americans-concerned-zika-will-spread-to-their-neighborhoods/; "More Than 4 in 10 Mistakenly Think Zika Is Fatal, Symptoms are Noticeable," March 10, 2016, http://www.annenbergpublicpolicycenter.org/more-than-4-in-10-mistakenly-think-zika-is-fatal-and-symptoms-are-noticeable/; "Who Does the Public Think Is Most at Risk from Zika?," April 8, 2016, Annenberg Public Policy Center, http://www.annenbergpublicpolicycenter.org/who-does-the-public-think-is-most-at-risk-from-zika/.

173 But in late 2009, I met: Donald G. McNeil Jr., "Flu Story: A Pregnant Woman's Ordeal," *New York Times*, Oct. 19, 2009, http://www.nytimes.com/2009/10/20/health/20pregnant.html.